A GENERAL MODEL OF LEGGED LOCOMOTION ON NATURAL TERRAIN

THE KLUWER INTERNATIONAL SERIES
IN ENGINEERING AND COMPUTER SCIENCE

ROBOTICS: VISION, MANIPULATION AND SENSORS

Consulting Editor: **Takeo Kanade**

PERCEPTUAL ORGANIZATION AND VISUAL RECOGNITION, D. Lowe
ISBN: 0-89838-172-X
ROBOT DYNAMICS ALGORITHMS, F. Featherstone
ISBN: 0-89838-230-0
THREE- DIMENSIONAL MACHINE VISION, T. Kanade (editor)
ISBN: 0-89838-188-6
KINEMATIC MODELING, IDENTIFICATION AND CONTROL OF ROBOT MA-
NIPULATORS, H.W. Stone
ISBN: 0-89838-237-8
OBJECT RECOGNITION USING VISION AND TOUCH, P. Allen
ISBN: 0-89838-245-9
INTEGRATION, COORDINATION AND CONTROL OF MULTI-SENSOR ROBOT
SYSTEMS, H.F. Durrant-Whyte
ISBN: 0-89838-247-5
MOTION UNDERSTANDING: Robot and Human Vision, W.N. Martin and J. K.
Aggrawal (editors)
ISBN: 0-89838-258-0
BAYESIAN MODELING OF UNCERTAINTY IN LOW-LEVEL VISION, R. Szeliski
ISBN 0-7923-9039-3
VISION AND NAVIGATION: THE CMU NAVLAB, C. Thorpe (editor)
ISBN 0-7923-9068-7
TASK-DIRECTED SENSOR FUSION AND PLANNING: *A Computational Approach,*
G. D. Hager
ISBN: 0-7923-9108-X
COMPUTER ANALYSIS OF VISUAL TEXTURES, F. Tomita and S. Tsuji
ISBN: 0-7923-9114-4
DATA FUSION FOR SENSORY INFORMATION PROCESSING SYSTEMS, J. Clark
and A. Yuille
ISBN: 0-7923-9120-9
PARALLEL ARCHITECTURES AND PARALLEL ALGORITHMS FOR INTEGRATED
VISION SYSTEMS, A.N. Choudhary, J. H. Patel
ISBN: 0-7923-9078-4
ROBOT MOTION PLANNING, J. Latombe
ISBN: 0-7923-9129-2
DYNAMIC ANALYSIS OF ROBOT MANIPULATORS: A Cartesian Tensor Approach,
C.A Balafoutis, R.V. Patel
ISBN: 07923-9145-4
PERTURBATION TECHNIQUES FOR FLEXIBLE MANIPULATORS: A. Fraser and
R. W. Daniel
ISBN: 0-7923-9162-4
COMPUTER AIDED MECHANICAL ASSEMBLY PLANNING: L. Homen de Mello and
S. Lee   ISBN: 0-7923-9205-1
INTELLIGENT ROBOTIC SYSTEMS FOR SPACE EXPLORATION: Alan A. Desrochers
ISBN: 0-7923-9197-7
MEASUREMENT OF IMAGE VELOCITY: David J. Fleet
ISBN: 0-7923-9198-5
DIRECTED SONAR SENSING FOR MOBILE ROBOT NAVIGATION: John J. Leonard,
Hugh F. Durrant-Whyte   ISBN: 0-7923-9242-6

A GENERAL MODEL OF LEGGED LOCOMOTION ON NATURAL TERRAIN

By:

David J. Manko
Westinghouse Electric Corporation

Foreword by:

William L. Whittaker
Carnegie Mellon University

KLUWER ACADEMIC PUBLISHERS
BOSTON / LONDON / DORDRECHT

Distributors for North America:
Kluwer Academic Publishers
101 Philip Drive
Assinippi Park
Norwell, Massachusetts 02061 USA

Distributors for all other countries:
Kluwer Academic Publishers Group
Distribution Centre
Post Office Box 322
3300 AH Dordrecht, THE NETHERLANDS

Library of Congress Cataloging-in-Publication Data

Manko, David J., 1957-
　A general model of legged locomotion on natural terrain / by David J. Manko; foreword by William L. Whittaker.
　　p. cm. -- (The Kluwer international series in engineering and computer science. Robotics)
　　Includes bibliographical references (p.　) and index.
　　ISBN 0-7923-9247-7
　　1. Robotics. 2. Artificial legs. 3. Mobile robots. I. Title. II. Series.

TJ211.M355　1992
629.8'92--dc20   92-13547
                                                      CIP

Copyright © 1992 by Kluwer Academic Publishers

All rights reserved. No part of this publication may be reproduced, stored in a retrieval system or transmitted in any form or by any means, mechanical, photo-copying, recording, or otherwise, without the prior written permission of the publisher, Kluwer Academic Publishers, 101 Philip Drive, Assinippi Park, Norwell, Massachusetts 02061.

Printed on acid-free paper.

Printed in the United States of America

Contents

List of Figures vii

List of Tables xi

Foreword x

1 Introduction 1
  1.1 Motivation 1
  1.2 Problem Statement 2
  1.3 Monograph Overview 3

2 Background 5
  2.1 Existing Models of Legged Locomotion 5
  2.2 Dynamic Formulation 6
    2.2.1 Mechanism Dynamics 7
    2.2.2 Constraint Modeling 9
  2.3 Foot-Soil Interactions 10
    2.3.1 Vertical Force-Deflection Relationships 10
    2.3.2 Lateral Force-Deflection Relationships 13
  2.4 Joint Modeling 13
    2.4.1 Coulomb Friction 14
    2.4.2 Viscous Damping 15

3 Legged Locomotion Model 17
  3.1 Generalized Coordinates 17
  3.2 Dynamic Formulation 18
  3.3 Foot-Soil Interactions 20
    3.3.1 Vertical Loading 21
    3.3.2 Lateral Loading 21
    3.3.3 Rotational Loads 25
    3.3.4 Slope Conditions 25
    3.3.5 Approximated Structural Compliance 25
  3.4 Joint Damping and Backdriving 26

4	**Solution Procedures**	**33**
	4.1 Numerical Solution Scheme	33
	4.2 Iterative Methods	35
	4.3 Implementation	37

5	**Application to a Prototype Walking Machine**	**39**
	5.1 AMBLER Design	39
	5.2 Geometric Representation	42
	5.3 Physical Parameters	45
	5.4 Specialization of Dynamic Equations	45
	5.5 Massless Leg Model	48

6	**Verification Studies**	**51**
	6.1 Model and Solution Parameters	52
	6.2 Flat Settlement Examples	54
	6.3 Sloped Settlement Examples	59
	6.4 Body Move Examples	63
	6.5 Leg Recovery Example	68
	6.6 Computational Performance	70

7	**Model Applications**	**73**
	7.1 Gait Cycle Simulations	73
	7.2 Body Leveling Simulations	75
	7.3 Example of Model Generality	84

8	**Summary**	**91**
	8.1 Summary	91
	8.2 Future Research	93
	8.3 Perspectives	93

A	**Vertical Foot-Soil Experiments**	**97**

B	**Perturbation Parameters**	**103**

C	**Inertial Properties for the AMBLER Massless Leg Model**	**105**
	C.1 Reallocation of Inner Link Properties	106
	C.2 Reallocation of Outer Link Properties	107
	C.3 Reallocation of Vertical Link Properties	107

Bibliography   109

Index   115

List of Figures

2.1	Coulomb Friction Models	15
3.1	Effects of Foot Forces on Body Rotations	20
3.2	Vertical Reloading of Two Soil Types	22
3.3	Lateral Foot Displacement Vector	22
3.4	Pseudo Reloading of the Lateral Force-Deflection Curve	23
3.5	Examples of Lateral Pseudo Reloading Curves	24
3.6	Foot Placement on a Sloped Surface	26
3.7	Augmented Foot-Soil Interactions	27
3.8	Joint Damping Test Configuration	27
3.9	Joint Damping Test Results	28
3.10	Joint Backdriving Test Configuration	29
3.11	Joint Backdriving Test Results	30
5.1	AMBLER Walking Machine Configuration	40
5.2	AMBLER Leg Design	40
5.3	AMBLER Recovering Leg Motions	41
5.4	AMBLER Coordinate Frames	42
5.5	Constant Transformation to Realign Coordinate Axes	44
5.6	AMBLER Structural Flexibilities	47
6.1	Force-Deflection Relationships for a Compact Sand	53
6.2	Mechanism Stance Used for Flat Settlement Examples	55
6.3	Flat Settlement Results for Case 1	56
6.4	Flat Settlement Results for Case 2	57
6.5	Flat Settlement Results for Case 3	58
6.6	Mechanism Posture Used for Sloped Settlement Examples	59
6.7	Sloped Settlement Results for Case 1	60
6.8	Sloped Settlement Results for Case 2	61
6.9	Sloped Settlement Results for Case 3	62
6.10	Desired Trajectory for Body Move Simulations	64
6.11	Body Move Results for Case 1	65
6.12	Body Move Results for Case 2	66
6.13	Body Move Results for Case 3	67

6.14	Leg Recovery Simulation Results	69
7.1	Mechanism Stances Used for Gait Cycle Simulations	74
7.2	Force-Deflection Relationships for a Loose Sand	76
7.3	Force-Deflection Relationships for Martian Soil	77
7.4	Results for Gait Cycle Simulation on a Compact Sand	78
7.5	Results for Gait Cycle Simulation on a Loose Sand	79
7.6	Results for Gait Cycle Simulation on Martian Soil	80
7.7	Mechanism Stance Used for Body Leveling Simulations	81
7.8	Body Leveling Results for a Set of Stable Gains	82
7.9	Body Leveling Results for a Set of Unstable Gains	83
7.10	Overlapping Walker Configuration	84
7.11	Overlapping Walker Coordinate Frames	85
7.12	Mechanism Posture Used for Overlapping Walker Simulations	86
7.13	Overlapping Walker Flat Settlement Results	88
7.14	Overlapping Walker Body Move Results	89
A.1	Measured Foot Force Components	98
A.2	Experimental Results for 0° Ground Slope	98
A.3	Experimental Results for 10° Ground Slope	99
A.4	Experimental Results for 20° Ground Slope	100
A.5	Experimental Results for 30° Ground Slope	101
A.6	Graphical Interpretation of Bilinear Equation Parameters	102

List of Tables

5.1	AMBLER Denavit-Hartenberg Parameters	43
5.2	AMBLER Physical Properties	46
5.3	AMBLER Joint Damping and Backdriving Parameters	47
5.4	Degenerate Partial Derivative Matrices	48
6.1	Recovering Leg Motions	68
6.2	Iterative Solution Timing Studies	71
7.1	Gait Cycle Mechanism Motions	75
7.2	Overlapping Walker Denavit-Hartenberg Parameters	85
7.3	Overlapping Walker Physical Properties	87
7.4	Overlapping Walker Joint Damping and Backdriving Parameters	90
B.1	Convergence Ratios for Different Sets of Perturbation Parameters	104

Foreword

Dynamic modeling is the fundamental building block for mechanism analysis, design, control and performance evaluation. One class of mechanism, legged machines, have multiple closed-chains established through intermittent ground contacts. Further, walking on natural terrain introduces non-linear system compliance in the forms of foot sinkage and slippage. Closed-chains constrain the possible motions of a mechanism while compliances affect the redistribution of forces throughout the system.

This research monograph, which is based on David Manko's dissertation research at Carnegie Mellon University, develops a dynamic mechanism model that characterizes indeterminate interactions of a closed-chain robot with its environment. The approach is applicable to any closed-chain mechanism with sufficient contact compliance although legged locomotion on natural terrain is chosen to illustrate the methodology. The modeling and solution procedures are general to all walking machine configurations, including bipeds, quadrupeds, beam-walkers and hopping machines.

The general approach is to identify and model the contact compliance, and then define a set of generalized coordinates that are advantageous for modeling the robot. Equations of motion are formulated which are applicable to all walker configurations through suitable definition of geometric parameters. The nature of the ground compliance is determined empirically from experimentation, and related to existing terramechanics models. Novel solution procedures are developed specifically for this class of problem that accommodate all combinations of imposed boundary conditions and these procedures have proven to be extremely robust.

This work develops a functional model of legged locomotion that incorporates, for the first time, non-conservative foot-soil interactions in a non-linear dynamic formulation. The model was applied to a prototype walking machine, and simulations generated significant insights into walking machine performance on natural terrain. The simulations are original and essential contributions to the design, evaluation and control of these complex robot systems. While posed in the context of walking machines, the approach has wider applicability to rolling locomotors, cooperating manipulators, multi-fingered hands, and prehensile agents. I believe that this monograph provides methods, insights and is a reference that will be useful to researchers in these areas.

Controlled closed-chain devices are emerging as an important class of capable mechanisms. Walkers, hands and multiple arms, which have been discounted for their complexity and control difficulties, are now maturing into viable robot forms. This work is a milestone toward the understanding and realization of these important robots forms of our future.

William L. Whittaker
Field Robotics Center
Carnegie Mellon University

Chapter 1

Introduction

Walking machines have the potential to navigate terrains that are inaccessible to wheeled locomotors. Advantages of legged locomotion include terrain-isolated propulsion, reduced soil work, a stable work platform, smooth body trajectories, and adaptability of gaits for obstacle avoidance. However, existing walking machines are research prototypes limited to navigating smooth, non-compliant surfaces. There is virtually no experience with autonomous, integrated systems that traverse rugged terrain featuring varying degrees of compliance. Walking on natural terrain imposes transitional foot contact loads, foot sinkage and slippage, and foot placements on rocks, uneven surfaces and slopes.

Research in this area requires a functional model of legged locomotion on natural terrain to be used for simulation studies, performance evaluations, and model-based control. Applications include development of mechanical configurations that minimize power consumption, maximize stability, and increase payload-to-weight ratio. In addition, advances in perception, planning, and fail-safe control are needed for fully autonomous operation. Gait selection and footfall planning are unique to legged locomotion and require significant modeling before legged mobility can be fully capable in unstructured terrains.

1.1 Motivation

Sufficient analysis of a walking mechanism is required to guarantee that power and performance goals can be attained because challenges in previous development of physical walkers - immature components, integration complexity, cost, scale and fabrication issues - limited the possible extent of earlier experimental investigation. A model of legged locomotion on natural terrain is required to develop control algorithms, gait planning and selection, and performance criteria for power consumption, stability margins and maximum grade of traversal. Estimates of mechanism force distributions during operation, which are necessary for mechanical design, can be established through model simulations. Also, control algorithms can be formulated and tested in a timely and less committing manner through model simulations compared to direct development on robotic hardware.

Locomotors are gravity stabilized and have a potential for tipover, which could be disastrous to a mission. On rough terrain, an autonomous walking machine must plan a path to the desired destination, while considering the potential for tipover during traversal. A functional model of legged locomotion on natural terrain is useful for proactive assessment of candidate paths and gait selections considering machine stability and power consumption.

Many walking machine configurations could conceivably operate on natural terrain, and it is desirable that the formulated legged locomotion model be sufficiently general to all designs. Also, successful simulations depend on the development of solution procedures which are appropriate for all possible applications of the model. A model of legged locomotion on natural terrain must apply to extreme events, such as accidental foot contacts, with fidelity and robustness that is transparent to a community of users from designers to power engineers to control scientists.

1.2 Problem Statement

Development of a functional model is essential for realization of legged locomotion on natural terrain. Although most walking machines operate with relatively small accelerations and, therefore, experience minimal inertial loadings, compliances in the robot and environment can produce significant time-dependent mechanism responses that must be evaluated with a dynamic formulation. Further, compliance in a closed-chain mechanism (e.g., a walking machine on compliant terrain) serves to redistribute forces throughout the system, a phenomenon which must be modeled in detail for the design and control of these systems. Existing legged locomotion models have not incorporated system compliance, particularly foot-soil interactions, in a full dynamic formulation.

A model of legged locomotion on natural terrain is formulated in this work that incorporates non-linear foot-soil interactions into a full dynamic formulation. Otherwise, link members and joints are considered to be ideally rigid, allowing a tractable number of degrees of freedom (dofs). Appropriate joint damping and backdriving models are identified through experimentation. Novel solution procedures are devised for this class of problem which are extremely robust and adaptable to all problem variations arising from different combinations of imposed boundary conditions. The modeling and solution procedures are general to all walking machine configurations including bipeds, quadrupeds, beam-walkers, and hopping machines.

1.3 Monograph Overview

Chapter 2 presents background relevant to modeling legged locomotion on natural terrain; existing models of legged locomotion, dynamic formulations, constraint modeling, representations of foot-soil interactions, and joint modeling are discussed. Formulation of the legged locomotion model including closed-chain mechanism dynamics, ground interactions and joint behavior is described in Chapter 3. Procedures are presented in Chapter 4 which efficiently calculate stable, accurate solutions in a novel manner, once the model has been formulated.

The legged locomotion model is applied to a prototype walking machine in Chapter 5 and simulations described in Chapter 6 are used to verify the modeling and solution procedures. The utility of the model is demonstrated through gait cycle and leveling control simulations of the prototype walking machine and model generality is illustrated by application of the model to an alternate configuration (Chapter 7). A summary of this monograph is provided in Chapter 8. The appendices detail foot-soil testing, numerical solution parameters, and development of a simplified massless leg model.

Chapter 2

Background

This monograph develops a model of legged locomotion on natural terrain that, for the first time, incorporates non-conservative foot-soil interactions into a full dynamic formulation. Hence, methods for deriving the equations of motion for robotic systems are essential building blocks. Further, legged machines have multiple closed-chains[1] established through ground contacts that must be represented in the dynamic formulation; approaches for modeling motion constraints are examined.

Walking imposes transitional foot contact loads, sinkage and slippage of feet, and foot placements on rocks, uneven terrain, and slopes. Therefore, characterization of foot contact in natural terrain is essential for prediction and control of off-road walking machines. Joint frictional and backdriving effects increase power consumption and can be problematic to a walker's control system. In order for the legged locomotion model to be useful for control system simulations and performance evaluations, joint modeling must adequately predict these effects. This chapter presents background relevant to these phenomena, which are essential to the modeling of legged locomotion on natural terrain.

2.1 Existing Models of Legged Locomotion

The earliest models of biped locomotion [1], [2], [3] only considered dynamics of the body while neglecting leg inertias and mechanism compliance. The formulation of [4] included leg inertias but system compliance was not addressed. These biped locomotion models were useful for calculating the mechanism response to a specified history of applied forces (i.e., forward dynamic model).

The non-compliant legged locomotion models in [5], [6] idealize a walking machine initially without ground contacts, then restore contact conditions in the course of analysis. Thus, the mechanism is transformed into a set of serial

---

[1] Closed path (i.e., a curve beginning and ending at the same point) in a structure around which forces are transmitted.

link manipulators with a common base (i.e., the body) for which the recursive Newton-Euler algorithm is applicable to each chain (or leg). The equations of motion are derived by considering the dynamic equations for each leg, equilibrium of the body, and constraint equations for each closed-chain.

The resulting system equations comprise a set of simultaneous linear algebraic equations, where the unknown quantities for the forward dynamic model [5] are the joint accelerations, body accelerations and constraint forces. In this formulation, the unknown accelerations and forces are assumed to remain constant during a timestep; displacements and velocities are calculated by explicit integration of the accelerations. The stepsizes required for an accurate solution are small and, therefore, more numerous than the stepsizes that are possible with implicit procedures.

Joint forces required to produce desired mechanism motions are calculated using inverse dynamic models; these models are obtained by substituting joint trajectories into the equations of motion. The inverse dynamic equations for a non-compliant, closed-chain mechanism are the same as the forward dynamic equations except that the constraint equations are no longer applicable because the specified joint motions identically satisfy the kinematic constraints. The unknown quantities for inverse dynamic models of non-compliant, legged locomotion [6] are the joint forces and constraint forces; these variables are underspecified since the constraint equations have been eliminated. Linear programming techniques were used to calculate optimized solutions that minimize input power while satisfying maximum limits on joint and traction forces.

Models of legged locomotion with compliant joints represented by triaxial springs are derived in [7]; a set of non-generalized coordinates were used to formulate the model without consideration of closed-chains established through ground contacts. The resulting model predicts kinematically inadmissible motions which violate the constraint equations that are required to enforce an admissible mechanism response. Vertical foot-soil interactions have been included in a static legged locomotion model of a hexapod walker with an alternating tripod gait [8], but this model is inappropriate for the indeterminant condition when more than three legs contact the ground.

2.2 Dynamic Formulation

Different methods for deriving the equations of motion for robotic systems are examined here in the context of walking machines. The discussion is limited to those methods that are applicable to rigid mechanisms (i.e., rigid members and joints) with compliant contacts because the legged locomotion model formulated in this work considers the mechanism to be rigid (with an option for lumping mechanism compliance at the contact points). Legged locomotion is characterized by multiple closed-chains established through ground contacts that affect the resulting mechanism motions. Approaches for modeling the constraints of closed-chain mechanisms are discussed.

2.2.1 Mechanism Dynamics

Lagrangian Dynamics Lagrange's equations of motion are specified as functions of the potential and kinetic energies of a body. As a result, interbody constraint forces do not require consideration because these forces produce no useful work (i.e., do not add to the energy of the system). The use of scalar energy expressions combined with eliminating the need to consider constraint forces permits relatively simple application of Lagrangian dynamics to complex configurations when compared to other methods that use vector quantities. Application of Lagrange's equations of motion to robotic systems using (4×4) homogeneous transformation matrices (for specifying the mechanism geometry) represents a formal method of formulating dynamic equations [31]; the derivation of this approach is commonplace and will not be reiterated here.

Lagrangian dynamics is shown in [9] to be inefficient relative to the number of additions and multiplications required [9] for inverse dynamic calculations. A recursive Lagrangian dynamic formulation is presented in that reference where the coefficients for the dynamic equations are calculated recursively, resulting in a reduced computational dependency. Mechanism closed-chains can be represented in a Lagrangian formulation with constraint equations and Lagrange multipliers (or equivalent constraint forces), or with a reduced set of generalized coordinates.

Newton-Euler Dynamics Alternately, the equations of motion for a mechanism can be derived by applying the Newton-Euler equations to each link [32]. The Newton-Euler equations account for all forces and moments acting on a link including constraint forces between links and any applied forces (i.e., joint forces, contact loadings, etc.). The system equations are obtained in any suitable inertial frame by assembling the individual link equations while eliminating constraint forces between links.

The Newton-Euler formulation requires consideration of force vectors (i.e., magnitude and direction), which combined with the need to eliminate constraint forces, makes this approach more difficult to apply to complex configurations compared with other methods that consider only scalar quantities (e.g., work or energy). Constraint forces resulting from closed-chains are transmitted through a mechanism affecting the interlink constraint forces. Since constraint forces between links require explicit consideration with the Newton-Euler formulation, closed-chains must be represented with constraint equations and Lagrange multipliers (or equivalent constraint forces).

A recursive Newton-Euler algorithm [10] is available for the efficient calculation of inverse dynamic solutions for tree structured mechanisms. The recursive approach calculates link velocities from the base to distal links and then calculates link forces from the tip (or end-effector) to the base. The joint forces required to produce the desired mechanism trajectory are obtained from the calculated link forces.

D'Alembert's Principle D'Alembert's Principle can be used to derive the equations of motion for a robotic system by defining all forces acting on a body which is undergoing an acceleration; the vector sum of forces results in equilibrium equations for a member. The link equilibrium equations are simultaneously solved, consistent with the boundary conditions, to obtain the dynamic response of the system. Closed-chains must be represented with constraint equations and Lagrange multipliers (or equivalent constraint forces) because interlink constraint forces must be explicitly defined.

Definition of all forces acting on a member is difficult because centrifugal and coriolis forces are not easily envisioned for inclusion in the equilibrium equations. As a result, centrifugal and coriolis effects are usually neglected when using D'Alembert's principle which limits application of the method to mechanisms that operate at relatively low speeds. Use of force vectors makes this approach more difficult to apply relative to other methods that consider only scalar quantities (e.g., work or energy).

Kane's Dynamics Kane's dynamical equations are yet another method used for formulation of the equations of motion for robotic systems [11]. Kane's dynamics is a subset of the more general class of methods known as Lagrange's form of D'Alembert's principle (or Lagrange's Principle) [12]. The essence of Kane's dynamics is to multiply the Newton-Euler equations with selected vectors (called partial velocities and partial angular velocities, which are non-dimensional quantities) to obtain scalar representations of the forces acting on a body. As a result of the dot product operation, interbody constraint forces are shown [49] to cancel during assembly of the dynamic equations and do not require consideration.

A set of generalized speeds is defined for a configuration and the linear and angular velocities of each member determine the partial velocity and partial angular velocity vectors used to multiply the Newton-Euler equations. Generalized speeds can be expressed as functions of the joint velocities, which is the same form as non-holonomic constraint equations (i.e., functions of joint velocities) representing closed-chains in a mechanism. As a result, a dependent set of generalized speeds is amenable to coordinate reduction through Gaussian elimination into an independent set of variables [33]. Therefore, constraint equations can be represented in Kane's dynamics through definition of an independent set of generalized speeds [13], Gaussian elimination of a dependent set of generalized speeds, or Lagrange multipliers and constraint equations.

A judicious choice of generalized speeds can simplify the resulting equations of motion as shown by the application of Kane's dynamics to the Stanford Arm [11]. Other advantages of this approach include the relative ease of scalar operations and direct accommodation of closed-chains in a mechanism by a suitable definition of generalized speeds. Application of Kane's dynamics requires extensive symbolic manipulations and the appropriate choice of generalized speeds is not always apparent, which are disadvantages of the method. Also, the resulting model is suitable only for the mechanism configuration under consideration.

2.2.2 Constraint Modeling

Closed-chains in a mechanism constrain the possible motions of that system. Holonomic constraints that characterize fixed-base manipulators can be expressed as functions of position variables while non-holonomic constraints that apply to wheeled mobile robots can only be defined in terms of velocities; constraint type can be determined as shown in [14]. Legged locomotion is characterized by holonomic constraints because the associated constraint equations can be expressed as functions of the joint positions [15]. The type of constraint can affect the choice of solution technique [34]; simpler conditions (and procedures) are associated with holonomically constrained systems.

Closed-chain mechanisms are most effectively modeled with generalized coordinates that produce a minimum number of dynamic equations from the outset. This approach eliminates the need to generate, then condense excess equations. A valid set of generalized coordinates must be an independent set of variables that represent all kinematically admissible mechanism motions [35]. Therefore, all possible mechanism motions can be uniquely expressed as functions of the generalized coordinates. General guidelines are not available for identifying sets of generalized coordinates because each configuration may require a different set of coordinates. The choice of coordinates is based primarily on the analyst's experience and insight.

Lagrange multipliers are commonly used for incorporating constraints into the equations of motion [16] for a mechanical system. The constraint equations (specified as velocity functions) are multiplied by Lagrange multipliers and the coefficients of each velocity variable are appended to the dynamic equation for that variable. (These additional terms represent constraint forces acting on the corresponding dof.) The equations of motion are composed of dynamic and constraint equations where the unknowns include the Lagrange multipliers in addition to the dependent set of variables.

Advantages of using Lagrange multipliers are that the method is formalized, suitable for all applications, and constraint forces are explicitly calculated. Disadvantages of the approach are the increased number of variables and equations compared to other methods that reduce the dependent set of variables. Also, equations of motion that have been formulated with Lagrange multipliers are a singular set of differential equations that require special solution procedures [36].

The remaining methods for incorporating constraints into the equations of motion involve reducing the dependent set of variables consistent with the constraint equations. The simplest approach is to apply Gaussian elimination to solve for a subset of coordinates as functions of the remaining variables. These expressions are back-substituted into the dynamic equations, thus reducing the set of variables. Gaussian elimination of position variables is possible only in relatively simple cases and solution singularities can occur because of constraint changes or the choice of eliminated coordinates [18].

More general approaches for reducing the original set of coordinates use orthogonal transformations characterized by the eigenvalues and eigenvectors of the constraint equations. The zero-eigenvalues theorem [17] defines a transformation matrix as a collection of independent eigenvectors associated with the zero eigenvalues of $A^T A$ where A is the coefficient matrix of the constraint equations. Another orthogonal transformation method results from application of singular value decomposition to the coefficient matrix [18] for definition of the transformation matrix from original to reduced coordinates. Coordinate reductions through orthogonal transformations are formal procedures that avoid the solution singularities associated with Gaussian elimination. The disadvantage of using these reduction techniques is the computational complexity of the required operations.

2.3 Foot-Soil Interactions

Soil loading characteristics of a walking machine foot pad are similar to the those observed during plate testing on soil. Walking machines are relatively slow moving (existing machines have a top speed of $8\,mph$) and do not propel soil mass under normal operation so that soil dynamic effects can be neglected in the representation of foot-soil interactions. Soil response to foot placement is fast relative to foundation settlement, which eliminates the need to consider consolidation due to reduction of pore water pressure over time.

Modeling foot-soil interactions requires force-deflection relationships for different loading and slope conditions of foot placement. The bulk of relevant technology is located in the terramechanics literature which is a discipline that predicts vehicle performance on natural terrain. A fundamental approach in terramechanics is to determine (experimentally or analytically) force-deflection characteristics of a representative soil sample loaded by plates of various sizes and shapes. This information is then extrapolated to predict vehicle performance.

Available force-deflection relationships for modeling vertical and combined vertical-lateral plate loading conditions are discussed below. No available references addressed the special cases such as rotational loading of a plate on soil under various combinations of vertical/lateral loads, placement on sloped surfaces, or penetration of variably oriented plates into soils. Significant research efforts are required to completely understand the complex phenomena of foot-soil interactions and the effect of foot design parameters such as sole composition and shape.

2.3.1 Vertical Force-Deflection Relationships

Empirical force-deflection relationships are used to characterize vertical plate loading on soils. The method is applied by fitting parameters of a chosen functional relationship to experimental load-deflection data. Variations in soil type (e.g., clay, sand) and soil properties with depth (non-homogeneous soils) are characterized by definition of parameters and choice of function types.

Background

The power equation (Equation 2.1) is commonly used to represent the vertical force-deflection behavior of a plate on a homogeneous, compact soil [19].

$$F = kz^n \qquad (2.1)$$

where F - vertical force,
z - vertical sinkage, and
k, n - constants.

This equation was modified to make it independent of test plate size and shape, but the same functional form (i.e., displacement raised to a power) was retained. An implicit assumption of the power equation is that data reduces to linear form when plotted on logarithmic axes, which is rarely the case. The non-linearity is the result of elastic soil response prior to soil failure and the modified power equation (Equation 2.2) was proposed to provide a better data fit.

$$F - F_o = kz^n \qquad (2.2)$$

where F_o - surface bearing capacity load.

The above modification removes the non-linearity attributed to elastic deformation (when plotted on logarithmic scale) and provides a better fit at higher sinkages. A drawback of this modification is the suppression of elastic response, which results in poor predictions at lower sinkages that are relevant to walking. Application of an exponential factor was proposed [20] to characterize the transition region from zero sinkage to full plastic deformation. The resulting exponential equation is

$$F = \left\{ 1 - \exp\left(\frac{-z}{z_t}\right) \right\} [F_o + kz^n] \qquad (2.3)$$

where z_t - transition parameter.

In effect, an elastic region is retrofit to the full plastic response curve.

The following equation has been used for predicting the vertical force-deflection response of clays and loose soils.

$$F = F_u \left[1 - \exp\left(\frac{-z}{k}\right) \right] \qquad (2.4)$$

where F_u - asymptotic load.

According to this equation, the vertical force tends to an asymptote (a characteristic of cohesive soils) and does not increase at higher displacements.

A compromise between the power and exponential equations, the bilinear equation (Equation 2.5) forms an exponential transition from zero sinkage to a linear plastic response [21]. Although the following equation contains constant, exponential and linear terms, it is traditionally called the bilinear equation.

$$F = F_i \{1 - \exp\left[-(k_o - k_f)z\right] + k_f z\} \quad (2.5)$$

where F_i - intercept of the plastic response line with the load axis, and
k_o, k_f - slope parameters.

The bilinear equation explicitly represents the elastic and plastic soil response but constrains the plastic response to be linear; a linear plastic response may be inappropriate for some soils at higher deformations.

The depth of soil affected by a vehicle (either wheeled or legged) is limited to approximately .6 m in the most severe cases [24]. At this limited soil depth, it is reasonable to consider, at most, two soil strata when considering non-homogeneous soil types. A weaker soil on top of a stronger base exhibits a response similar to homogeneous soil types and the force-deflection relationships already discussed are applicable. An exponent greater than 1 for the power-type equations or a plastic slope greater than the elastic slope for the bilinear equation characterizes soil stiffening, which can occur as stress waves reach deeper, firmer strata with increasing penetration.

A different behavior is experienced when a strong soil layer covers a weak soil such as muskeg [25]. The top layer fails by punching through to the weaker soil, which can sustain no appreciable load. Therefore, the interesting region of the force-deflection curve for such soils is for sinkages up to collapse of the top layer. The equations discussed above can be used to define the foot-soil response for these soils up to the failure load.

Dimensional analysis has been applied [41] to predict the vertical force-deflection response of a plate on soil. Because of formulation restrictions, the method has been applied only to purely frictional or purely cohesive soils. The resulting equations obtained from dimensional analysis are identical in form to the empirical power equation (refer to Equation 2.1).

The analytical method of predicting the vertical force-deflection response of a plate on soil was originally developed for calculation of foundation bearing capacity; this approach has been subsequently applied to terramechanics calculations [19]. The vertically loaded plate is assumed to sink to a depth where the applied load is just below the bearing capacity of the soil. Plate penetration is accompanied by plastic soil flow along an assumed failure surface. For a given soil type and plate size, the analytical equation is equivalent in form to Equation 2.2 which is the empirical power equation modified to remove the elastic response.

Repetitive vertical soil loading by flat plates was investigated by [21] for a remolded London clay and by [24] for dry loose sand. In both cases, the unloading portion of the curve (associated with lifting of a walking machine foot) follows a straight line with a slope equal to the slope at zero sinkage until no load is applied. Subsequent reloading increases along the line of initial slope to the point where the previous maximum has been reached. Further loading (past the previous maximum) follows the original force-deflection curve until the cycle is repeated.

2.3.2 Lateral Force-Deflection Relationships

Representations of lateral plate loading (under a combined vertical load) on soil are based on Coulomb's equation (Equation 2.6) which describes the failure envelope of a soil sample.

$$T_{max} = c + p \tan \phi \tag{2.6}$$

where T_{max} - maximum lateral force,
c - cohesion,
p - normal load, and
ϕ - angle of internal friction.

This equation defines the asymptote that bounds the lateral force (i.e., maximum) for a given normal load. Different transition factors have been used to describe the force-deflection curve from zero load and displacement to the maximum load defined by Coulomb's equation. The following exponential transition is used for plastic (i.e., non-brittle) soils [42].

$$T = T_{max} \left[1 - \exp\left(\frac{-j}{k}\right)\right] \tag{2.7}$$

where T - applied lateral force,
j - horizontal displacement, and
k - transition constant.

A characteristic of brittle soils (i.e., strong soil layer on top of a weaker layer) is a humped force-deflection curve for lateral loading where the shear stress at failure is momentarily exceeded. This type of response is similar to an underdamped vibration response and equations have been proposed [22], [25] based on this analogy.

All previous equations used to describe lateral force-deflection behavior implicitly assume the vertical force remains constant during the change in lateral loading. This assumption is shown to be valid [23] for moderate loads and deformations which categorizes legged locomotion on natural terrain. References could not be located in the published literature for special cases such as repetitive lateral loading of a plate on soil.

2.4 Joint Modeling

Frictional effects are experienced during any mechanical operation resulting in increased power consumption and possible control difficulties. Force control is commonly used for legged locomotion and stiction is particularly troublesome because the sudden transition (i.e., near singularity) from static to dynamic frictional forces at the onset of motion can create instabilities. In order for the legged locomotion model to be useful for control system simulations and performance evaluations, joint modeling must adequately predict frictional effects.

Friction has been identified as having many forms but it is typically modeled as coulomb friction and/or viscous damping, both of which are discussed below.

Joint backdriving is unique to closed-chain robotic mechanisms where an unpowered joint is forced to follow overall mechanism motions. Instead of the actuator transmitting power through the drivetrain and moving the joint, the link motions "backdrive" the drivetrain and actuator. This backdriving motion must overcome the typically high gear ratio of the joint drivetrain resulting in potentially high power consumption. If the imposed force does not exceed the force required to backdrive the joint, mechanism binding or foot slippage for a walking machine could occur. Joint backdriving effects are peculiar to the joint design and experimental results are used to define an appropriate model of this phenomenon.

2.4.1 Coulomb Friction

Coulomb friction is a dissipative force between two bodies that are in relative motion (dynamic) or pending relative motion (static). The friction force opposes the motion, and its magnitude is proportional to the normal force between bodies; the coefficients of proportionality are the coulomb friction coefficients. For two contacting bodies, the static coulomb friction force (i.e., stiction) must be overcome before motion can occur. Once relative motion begins, the minimum force required to continue the motion is the dynamic coulomb friction force. The magnitude of the dynamic coulomb friction is assumed to be independent of the relative velocity between moving bodies.

Standard coulomb friction models are shown in Figure 2.1. The model of Figure 2.1.a has a discontinuity at zero velocity where the frictional force transitions instantaneously from the higher stiction value to a lower dynamic friction force. This discontinuity can cause instabilities (i.e., lack of convergence) in a numerical solution. Constraint addition-deletion has been used [26] to model this phenomenon where constraint equations and Lagrange multipliers maintain the joint in a fixed position (i.e., constrained) until the stiction force is exceeded. When this occurs, the joint constraints are removed from the equations of motion and the frictional force is reset to the lower dynamic friction value. The constraints and higher stiction force are reinstated if the joint velocity drops to zero.

Different smoothing transitions are used [27], [28] to approximate the discontinuity from static to dynamic friction forces, thus accommodating numerical solution of the equations of motion; two examples are shown in Figures 2.1.b and 2.1.c. There remains a singularity at zero velocity (i.e., infinite slope of the friction force curve) that could prevent solution convergence for stiff systems of differential equations [30]. The approximate coulomb friction model of Figure 2.1.d was suggested [29] where the vertical line at zero velocity is replaced by a straight line with finite slope. The slope of the line can be increased to improve the accuracy of the representation at the expense of smaller timesteps and increased computational effort.

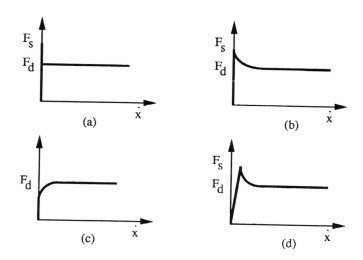

Figure 2.1: Coulomb Friction Models

2.4.2 Viscous Damping

The friction force at a joint has been observed to be a function of joint velocity and viscous damping has been used to model this phenomenon. Viscous damping is often a non-linear function of the joint velocity, but a linear response is usually assumed for simplicity or lack of detailed information on the physical behavior involved. Similarly, viscous damping is dependent on the normal force between bodies but this dependence is typically neglected. When viscous damping and coulomb friction occur together, they are superimposed to produce a composite damping response curve.

Chapter 3

Legged Locomotion Model

A general model of legged locomotion on natural terrain incorporating closed-chain mechanism dynamics, ground interactions and joint behavior is developed in this chapter. The model is unique for its representation of non-linear foot-soil interactions which are imposed as external forces acting on the mechanism. The modeling approach requires definition of generalized coordinates for legged locomotion and the equations of motion are derived using Lagrangian dynamics. Appropriate representations of foot-soil interactions, joint damping and backdriving are experimentally calibrated and incorporated into the model.

3.1 Generalized Coordinates

Formulation of any mechanism model requires selection of variables which describe the mechanism motions. The choice of coordinate variables for a legged locomotion model is complicated by the existence of multiple closed-chains which are repeatedly established and broken through intermittent ground contacts. Some representations of closed-chains require specification of constraint equations, thus producing a number of system equations and variables beyond a minimal set. Closed-chains of a walking mechanism are most effectively represented by generalized coordinates resulting in a minimum number of dynamic equations, thus increasing the computational efficiency.

A valid set of generalized coordinates must be an independent set of variables that represent all kinematically admissible mechanism motions. There are many sets of generalized coordinates suitable for modeling legged locomotion, each with associated advantages and disadvantages. The set of generalized coordinates consisting of 6 body dofs plus all appendage joint motions is discovered for modeling legged locomotion on natural terrain; coordinate sets for two configurations are shown in Figures 5.4 and 7.11. These coordinates are similar to those used in [54] to model a robot with a floating base where the body of a walking machine corresponds to the mechanism base.

The advantage of this set of generalized coordinates is its applicability to any walking machine configuration, and its consistency with homogeneous transformation matrices (i.e., link motions are expressed as functions of the joint motions). Closed-chains do not require explicit consideration since ground contacts are not implied with the generalized coordinate representation. (A model formulated using constraint equations would require a redefinition of the dynamic equations each time a ground contact is established or broken.) This approach is ideally suited for simulating tipover, freefall and foot placement/liftoff because varying numbers of ground contacting feet are accommodated without reformulation of the model.

Use of the above set of generalized coordinates assumes that modeled foot-soil interactions may resist, but do not absolutely prevent any motions of a ground contacting foot. Hence, it is essential to allocate some system compliance to the foot-soil interface. The coordinates are kinematically admissible because a variation of each dof is permissible while all other dofs are held fixed, demonstrating the independence of the coordinates. For example, varying the vertical body position causes increased vertical sinkage or liftoff of all feet. The foot-soil interaction models described in Section 3.3 do not prevent any motions of a ground contacting foot and these models are consistent with the chosen set of generalized coordinates.

3.2 Dynamic Formulation

The system equations for models of legged locomotion on natural terrain could be derived using alternate methods, but with corresponding disadvantages. Available mechanism dynamic simulation programs require significant modifications to incorporate non-conservative foot-soil interactions and intermittent ground contacts into a model. Application of Newton-Euler dynamics requires elimination of interbody constraint forces and symbolic manipulation of vector quantities, which is prohibitive for this class of problem. Kane's dynamical equations have the deficiencies that extensive symbolic operations are required during formulation and the resulting model is specific only to the mechanical configuration being considered.

Alternately, Lagrangian dynamics used in conjunction with homogeneous transformation matrices is a formal approach well suited for systematic application to complex configurations. Further, all walking machine configurations can be modeled simply by specifying geometric parameters without reformulation of the underlying methodology. This versatility is important for a prototyping environment where configuration changes such as adding appendages, evolving dimensions and changing types of joints are expected.

Both the recursive and non-recursive Lagrangian formulations provide the advantages of adaptability and systematic application. The non-recursive formulation is used to derive the equations of motion for legged locomotion on natural terrain because the resulting equations can be partitioned into state and non-state contributions as required by explicit solution schemes (e.g., explicit 2-stage Runge-Kutta). (Equations derived using the recursive formulation cannot be partitioned in this manner.) Therefore, the equations of motion derived using non-recursive Lagrangian dynamics are more general and solutions can be calculated by both implicit and explicit numerical procedures potentially reducing the computational effort. An anticipated drawback using the non-recursive Lagrangian approach is an increased computational effort (relative to the recursive method) required to evaluate the equations of motion but the resulting model generality outweighs this deficiency.

The generalized coordinates identified in Section 3.1 define the system state while the roll-pitch-yaw convention describes body rotations. The following equations of motion are derived for modeling legged locomotion on natural terrain (assuming 3 joints per leg and a variable number of legs) where q_j and F_j are the position and generalized force, respectively, for coordinate j.

Body Equations $\rightarrow$ j=1-6

$$F_j = \sum_{k=1}^{6} (D_{jk})_b \ddot{q}_k + \sum_{n=1}^{nlegs} \sum_{k=1}^{9} (D_{jk})_n \ddot{q}_{k^*} + \sum_{k=1}^{6} \sum_{m=1}^{6} (H_{jkm})_b \dot{q}_k \dot{q}_m$$
$$+ \sum_{n=1}^{nlegs} \sum_{k=1}^{9} \sum_{m=1}^{9} (H_{jkm})_n \dot{q}_{k^*} \dot{q}_{m^*} + (G_j)_b + \sum_{n=1}^{nlegs} (G_j)_n$$

Joint Equations $\rightarrow$ j=7-9 which gives j*=7-24 for n=1-6

$$F_{j^*} = \sum_{k=1}^{9} (D_{jk})_n \ddot{q}_{k^*} + \sum_{k=1}^{9} \sum_{m=1}^{9} (H_{jkm})_n \dot{q}_{k^*} \dot{q}_{m^*} + (G_j)_n \qquad (3.1)$$

$(D_{jk})_b = Tr\{U_{6k} J_6 U_{6j}^T\}$ $\qquad (D_{jk})_n = \sum_{i=j,k,7}^{9} Tr\{U_{ik} J_i U_{ij}^T\}$

$(H_{jkm})_b = Tr\{U_{6km} J_6 U_{6j}^T\}$ $\qquad (H_{jkm})_n = \sum_{i=j,k,m,7}^{9} Tr\{U_{ikm} J_i U_{ij}^T\}$

$(G_j)_b = -m_6 g^T U_{6j} \bar{r}_6$ $\qquad (G_j)_n = -\sum_{i=j,7}^{9} m_i g^T U_{ij} \bar{r}_i$

$(\ )^* = (\ )$ for $(\ ) = 1-6$

$(\ )^* = 3(n-1) + (\ )$ for $(\ ) = 7-9$

The above equations are partitioned into terms describing contributions from the body and leg dofs (i.e., coefficients with subscripts b and n, respectively) so that configuration changes (e.g., number of legs, leg to body connection, joint type) are easily incorporated into the model. Also, these equations

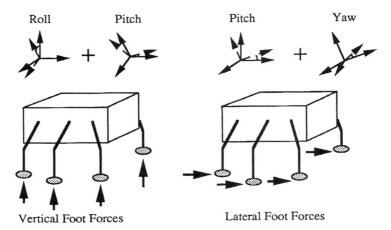

Figure 3.1: Effects of Foot Forces on Body Rotations

are amenable to parallel computation because leg contributions may be independently calculated. The generalized force, F_j, for each dynamic equation is composed of an applied force (actuated leg joints only), damping or backdrive force (leg joints only) and a force resulting from foot-soil interactions (all dofs). Foot forces are treated as external forces which are transformed into the body and joint coordinate frames for subsequent incorporation into the equations of motion.

3.3 Foot-Soil Interactions

Modeling foot-soil interactions requires force-deflection relationships for different loading conditions on flat and sloped surfaces. Consideration of vertical soil loading is required because a walking machine's weight is equilibrated by vertical foot forces. All walking machines are propelled by the generation of traction forces (i.e., lateral loads) between a foot and the soil. Therefore, lateral soil loading must be considered. Also, vertical foot forces generate roll and pitch rotations of the body while lateral foot forces create pitch and yaw body rotations as shown in Figure 3.1.

The walking machine leg and ankle designs affect the extent of rotational foot-soil loading as discussed below. Foot placement on sloped surfaces is inevitable, but not understood. Thus, approximations are necessary in the modeling of this phenomenon. Foot-soil interactions can be augmented with additional compliance that is representative of the mechanism's structural flexibility.

Legged Locomotion Model

3.3.1 Vertical Loading

The maximum vertical foot force applied by any walking machine built to date is approximately 7000 N. Assuming a 30 cm diameter foot pad, this load results in a vertical sinkage of $2 - 5\,cm$ on a typical soil. Since these sinkages are in the transition region from elastic to plastic behavior, the required foot-soil interaction model must accurately represent this portion of the load-deflection curve.

The bilinear equation (Equation 2.5) is used for modeling vertical foot-soil interactions.

$$F = F_i \left\{ 1 - \exp\left[-(k_o - k_f)\, z \right] + k_f z \right\}$$

This equation has two parameters to define the transition region (as opposed to one parameter for other equations), which provides a more accurate representation of this critical region. The bilinear equation is partially qualified for modeling vertical foot-soil interactions by the experiments and results described in Appendix A. A graphical interpretation of the bilinear equation and its parameters is given in Figure A.6.

Restricting the plastic response to be linear by using the bilinear equation is acceptable because expected sinkages are smaller than the range of sinkages where non-linear plastic response is observed. Degenerate cases of an extremely stiff base (e.g., rock) or a very weak base are modeled by adjusting the slope parameters of the bilinear equation. A weak soil over a strong base (e.g., a two layer stratified soil) is represented by a shallow initial slope and a steeper final slope of the curve.

Unloading and subsequent reloading of a foot in the vertical direction are modeled by straight lines with a slope equal to the initial or final slope of the bilinear equation, ($F_i k_o$ or $F_i k_f$ respectively), whichever is greater, until the previous maximum load has been reached. Further loading past the maximum previous load follows the original force-deflection curve. Modeling of vertical reloading for two soil types ($k_o > k_f$ and $k_o < k_f$) are shown in Figure 3.2. For an elastic soil type (i.e., $k_o = k_f$), the reloading model follows the elastic response curve.

3.3.2 Lateral Loading

Lateral foot-soil interactions are modeled with Equation 2.7 which describes an exponential transition to Coulomb's equation.

$$T = T_{max} \left[1 - \exp\left(\frac{-j}{k} \right) \right]$$

The only available alternatives are those equations analogous to an underdamped vibration response, which were intended for brittle soils. It is assumed that only plastic soils (i.e., non-brittle) will be encountered since this category accounts for the largest percentage of soils.

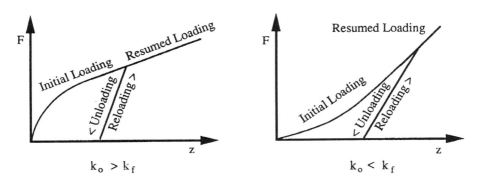

Figure 3.2: Vertical Reloading of Two Soil Types

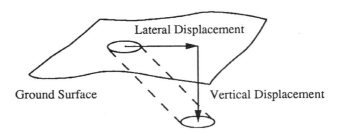

Figure 3.3: Lateral Foot Displacement Vector

Coulomb's equation defines the maximum lateral force assuming a constant vertical load and rectilinear motions. Vertical foot forces vary continuously during walking machine operations and foot motions are not constrained along a straight line. Therefore, application of Coulomb's equation for modeling lateral foot-soil interactions must consider a changing vertical load and unconstrained motions.

The lateral foot displacement vector is the projection of the total foot displacement vector (which connects the points of initial ground contact and current foot position) onto the ground plane as shown in Figure 3.3. For this work, the lateral displacement vector is decomposed into two orthogonal components (according to analyst specified direction vectors) whose orientations remain constant during a simulation. The magnitudes of the orthogonal vector components determine two lateral foot displacements and separate lateral soil loading events are considered. Therefore, total lateral foot motions are decomposed into two orthogonal, rectilinear motions and Coulomb's equation is applied individually to each lateral displacement.

This lateral foot-soil modeling approach does not consider coupling of the

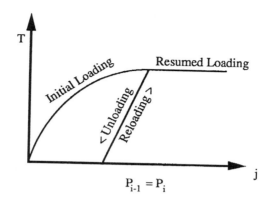

Figure 3.4: Pseudo Reloading of the Lateral Force-Deflection Curve

two orthogonal motions which is a phenomenon that has not been researched. The decoupled representation is most appropriate for simulating mechanism motions that generate traction forces having relatively constant orientations (e.g., straight line body motions). In these cases, the analyst defines direction vectors that decompose the total lateral displacement vector into primary and secondary components (i.e., parallel and normal to the traction forces, respectively). The secondary lateral foot forces are typically an order of magnitude smaller and coupling of the two lateral soil loadings is inconsequential.

The effect of a changing vertical load on the lateral force-deflection relationships is modeled by idealizing the transition between timesteps as unloading to a lateral displacement corresponding to zero lateral force, and reloading to the current displacement and vertical force (Figure 3.4). The unloading follows a straight line with a slope equal to the initial slope of the previous load-deflection curve until a zero lateral force is obtained. Reloading follows a straight line with a slope equal to the initial slope of the current load-deflection curve until the appropriate lateral displacement or current load-deflection curve (whichever comes first) is reached. In the latter case, the load-deflection curve is then followed to the current displacement.

The difference in slopes for the pseudo loading-unloading cycle is insignificant because the change in vertical load (the vertical load defines the initial slope) is small between timesteps. Applications of Coulomb's equation with an exponential transition for modeling lateral foot-soil interactions having various combinations of vertical loads are shown in Figure 3.5. (Vertical load changes between timesteps have been exaggerated for illustrative purposes.) This modeling approach directly accommodates repetitive lateral foot-soil loading. Significant research into foot-soil interactions must be completed before a general representation of lateral foot motions can be formulated.

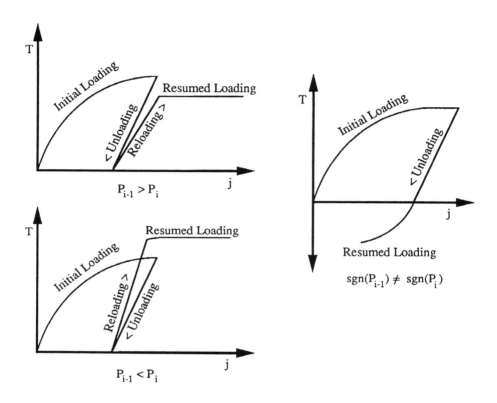

Figure 3.5: Examples of Lateral Pseudo Reloading Curves

3.3.3 Rotational Loads

Foot-soil rotations can occur depending on the walking mechanism's leg and ankle design. Biped locomotors commonly have a powered ankle joint intended to assist propulsion by generating foot-terrain moment loading. Indoor walking machine configurations capitalize on the relative non-compliance of their terrain by reducing foot size because foot sinkage is negligible and leg weaving is simplified. As a result, foot-terrain interactions for indoor walkers can be idealized as point contacts and even large foot rotations do not produce rotational loading.

Outdoor walking machines require larger footpads to prevent excessive sinkages on compliant terrains such as sea floors, sand, snow, mud, and planetary dust. Typically, if a leg configuration requires large ankle rotations, then the corresponding ankle is designed to minimize soil deformations. The pantograph leg design for the OSU Adaptive Suspension Vehicle (ASV) subtends large pitch motions [48] and a slaved ankle joint is used to maintain the foot level with the body. Similarly, the AMBLER orthogonal legged walker has a torsionally released ankle [47] to eliminate foot-soil rotations about the vertical leg axis during propulsion.

Rotations about constrained or slaved foot axes are produced by tilting of the body since foot and body orientations are equivalent. The prototype AMBLER walker is designed for a maximum body tilt of 5° in Earth gravity which produces negligible moment loadings on feet contacting level terrain. Rotational foot-soil loadings are neglected in this work because the leg design of the modeled AMBLER configuration has a torsionally released ankle joint and rotations about constrained foot axes are small. For configurations that generate significant rotational foot-soil loadings (i.e., biped locomotors), appropriate force-deflection relationships must be specified.

3.3.4 Slope Conditions

Foot placement on a sloped surface is a complex phenomenon for which no quantitative model is available (either empirical or analytical). The approximation is made here that foot placement on sloped surfaces is decomposed into components that are normal and tangent to the surface, respectively (Figure 3.6). These displacement components are then represented by vertical and lateral force-deflection relationships. There is no consideration of moment loading due to sloped foot-soil contact, an assumption which is justified, in part, by the relatively small foot dimensions of the modeled AMBLER configuration [47]. Cohesionless soils with unstable boundaries (e.g., dry sand) are not adequately modeled with the component approach because of penetration during loading.

3.3.5 Approximated Structural Compliance

The relative magnitude of structural flexibility versus soil compliance is dependent on the mechanism design and terrain type; all structures deform to

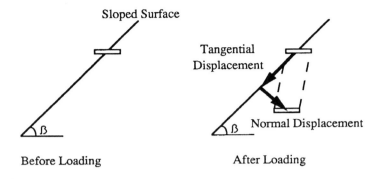

Figure 3.6: Foot Placement on a Sloped Surface

some extent. The current mechanism model considers all links and joints to be rigid, but foot-soil interactions can be augmented with additional compliance to approximate structural flexibility. Incorporation of structural compliance is essential for simulations on a rigid subgrade because the set of generalized coordinates specified in Section 3.1 would degenerate if foot motions were absolutely prevented.

Standard structural analysis techniques are used to calculate influence coefficients for a typically configured leg (i.e., leg joints in their most common positions). This information is used to determine additional deflections (both vertical and lateral) that are combined with the foot-soil data as shown in Figure 3.7. Finally, the foot-soil interaction equations are curve-fit to this modified data. In effect, a composite compliance is formulated for two springs in series (i.e., structural and terrain compliances). The maximum lateral foot-soil interaction force is unaffected by the added structural compliance; only the transition to the asymptote (parameter k in Equation 2.7) is affected.

3.4 Joint Damping and Backdriving

Joint damping and backdriving characteristics are dependent on the mechanism configuration and analytical models are not available for representation of these phenomena. Experiments were conducted with the Overlapping Walker Single Leg Testbed [52] to determine appropriate models of joint damping and backdriving. The leg was fully extended (i.e., straight elbow joint) to maximize the shoulder joint moment loading (increasing friction) and the foot was lifted off the ground for the damping tests as shown in Figure 3.8. A small velocity command was specified for the shoulder joint while maintaining fixed elbow and vertical joint positions. A build-up of velocity error resulted in the application of an increasing joint force until the shoulder joint started moving. Actuator current and encoder positions versus time were recorded throughout each of

Legged Locomotion Model 27

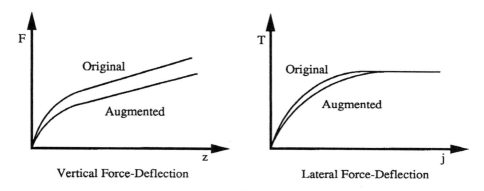

Figure 3.7: Augmented Foot-Soil Interactions

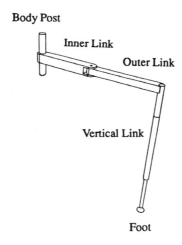

Figure 3.8: Joint Damping Test Configuration

3 tests. Results of a representative test, with data converted to torques and angles, are given in Figure 3.9.

A buildup of torque prior to any joint motion (i.e., stiction) is evident in the above results but the torque required to begin the motion is less than the magnitude required to sustain the motion. It appears that coulomb damping effects are greater than stiction effects for this mechanism and joint design. Therefore, the following equation is formulated for modeling damping effects of all leg joints.

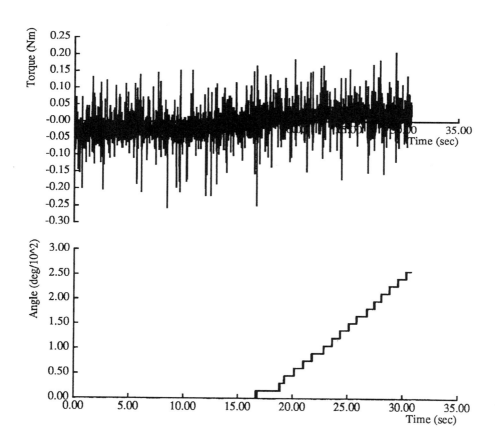

Figure 3.9: Joint Damping Test Results

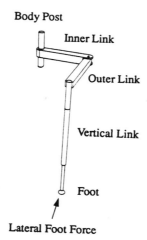

Figure 3.10: Joint Backdriving Test Configuration

$$F_d = F_{cf}\left[1 - \exp\left(-c_d \dot{q}\right)\right] \quad (3.2)$$

where F_d - joint damping force,
F_{cf} - coulomb friction value,
c_d - transition coefficient, and
$\dot{q}$ - joint velocity.

Only coulomb damping is considered with the above equation because joints of a walking machine move at relatively low velocities and viscous friction effects are not fully developed. The multiple force values for zero joint velocity that characterize stiction are not conducive for numerical solution (non-convergence could result) so the above equation approximates this phenomenon with an exponential transition to the steady state value. The transition rate can be varied with a quicker transition more closely representing stiction at the cost of smaller solution stepsizes. The exponential ramping provides a smoother transition and, therefore, affords more solution robustness than a linear approximation.

Joint backdriving was investigated using the Overlapping Walker test leg configured as shown in Figure 3.10. A lateral foot force was applied while elbow and vertical joint positions were kept fixed. Lateral foot forces and shoulder joint encoder positions versus time were recorded for each of 3 tests. Results of a representative test where the foot forces and encoder counts have been converted to torques and angles, respectively, are shown in Figure 3.11.

A build-up of torque occurs prior to any joint motion which is indicative of stiction. The stiction value (i.e., magnitude of torque at which motion begins) of 2.3 Nm is far less than the steady state value (60 Nm) measured during mechanical assembly for backdriving the shoulder joint power train. Therefore, the following equation is formulated for modeling backdriving effects of all leg joints.

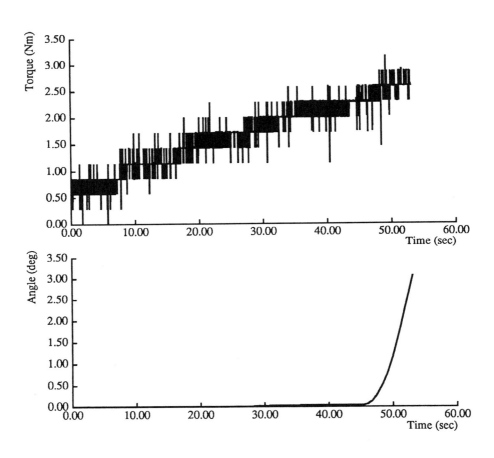

Figure 3.11: Joint Backdriving Test Results

Legged Locomotion Model

$$F_b = F_{ab}\left[1 - \exp\left(-c_b \dot{q}\right)\right] \tag{3.3}$$

where F_b - joint backdriving force,
F_{ab} - asymptotic backdriving force,
c_b - transition coefficient, and
$\dot{q}$ - joint velocity.

The reasoning for the above equation is the same as the reasoning discussed for the joint damping model.

Chapter 4

Solution Procedures

Procedures used to calculate solutions for a model of legged locomotion on natural terrain are discussed in this chapter. The equations of motion are solved in their original form for efficiency and adaptability to combinations of imposed boundary conditions (i.e., specified joint forces and trajectories). An Euler predictor-corrector scheme is applied to the differential equations and the estimated local error is used for global error control and variable stepsize definition. The non-linear algebraic equations obtained by discretizing the equations of motion are solved with a combination of full Newton, altered Newton and chord iterative methods. A difference approximation is used to calculate the system Jacobian because unloading characteristics and intermittent contact of the foot-soil modeling preclude symbolic evaluation. Relevant details of the solution implementation are described.

4.1 Numerical Solution Scheme

The equations of motion for a model of legged locomotion on natural terrain (Equation 3.1) are a set of ordinary differential equations (ODE's) prior to the application of any boundary conditions. If only joint force histories (i.e., F_j in Equation 3.1) are specified for a simulation, the equations of motion remain as ODE's and can be solved accordingly. When joint trajectories (i.e., q_i, $\dot{q}_i$ and $\ddot{q}_i$ in Equation 3.1) are specified for the model (e.g., gait simulation), the equations of motion are converted to an uncoupled set of differential/algebraic equations (DAE's) [34]. The differential equations corresponding to the joint variables with specified motions can be reduced to algebraic equations after substitution of the defined trajectories into the original equations of motion. The dynamic equations for all other variables having known force histories (i.e., unspecified motions) remain as differential equations. Solutions for an uncoupled set of DAE's can be calculated by first solving the differential equations to obtain the unknown mechanism motions and substituting these motions into the remaining equations to determine the joint forces required to produce the overall mechanism motions.

The differential equations for a model of legged locomotion on natural terrain are "stiff" for intervals of the mechanism response. A set of stiff equations is characterized by a relatively large difference between the smallest and largest time constants for the system. A stiff interval occurs for legged locomotion when impulsive foot loadings produce relatively high frequency mechanism responses that are superimposed upon lower frequency rigid body motions.

Appropriate algorithms for solving stiff differential equations must have the characteristic of stiff stability [43] so that reasonable stepsizes can be used. All stiffly stable methods are implicit, thus requiring an iterative solution to a set of non-linear algebraic equations (obtained by discretizing the differential equations) at each timestep. Numerical techniques exist [37] for switching between stiff and non-stiff solution algorithms (potentially reducing computational effort) that monitor the eigenvalues of the Jacobian. In this work, a stiffly stable solution algorithm is used throughout an entire simulation, because solution switching jeopardizes stability while necessitating additional operations (e.g., eigenvalue calculations) that reduce the potential advantage.

Existing ODE solution software assumes the equations are specified in a standard first order form. The second order equations of motion for a mechanical system must be transformed into an equivalent first order form for subsequent solution with these programs. The required transformation is dependent on the original form of the equations.

Calculating solutions for legged locomotion on natural terrain using available ODE packages requires a transformation of the equations of motion (after substitution of any specified joint trajectories) into standard DAE form [36] and the subset of differential equations converted to first order form. A specialized transformation is needed for each problem variation to obtain the required form of the equations. In this work, a special purpose solution method is applied to the original equations which does not require reordering or transformations. Furthermore, it is more efficient to solve the differential equations in their original form [44].

The implicit Euler formula used to discretize the differential equations is

$$\dot{\mathbf{y}}^{(i)} = \frac{\mathbf{y}^{(i)} - \mathbf{y}^{(i-1)}}{h} \tag{4.1}$$

where $\mathbf{y}^{(i)}$ - solution vector for timestep i, and
h - stepsize.

This formula and the associated algorithm are unconditionally stable [38] which produces a robust solution scheme. This is important for simulation of a walking machine experiencing impulsive loading (i.e., near discontinuities) from foot placement and soil breakout. The implicit Euler formula is a first order method with relatively poor accuracy from a numerical analysis standpoint, but the stability and ease of application compensate for the increased number of timesteps required to achieve a desired level of solution accuracy.

Solution Procedures

Application of an implicit algorithm requires a solution estimate at the beginning of each timestep. The explicit Euler formula (Equation 4.2) is used for solution prediction.

$$\dot{\mathbf{y}}_i = \frac{\mathbf{y}^{(i+1)} - \mathbf{y}^{(i)}}{h} \quad (4.2)$$

The Euler predictor-corrector solution scheme (i.e., combination of explicit and implicit formulas) provides an inexpensive local error estimate which is necessary for global error and automated stepsize control.

The estimated local error using the Euler predictor-corrector difference [38] is

$$d^{(i)} = \frac{1}{2} \left| \mathbf{y}_n^{(i)} - \mathbf{y}_0^{(i)} \right| \quad (4.3)$$

where $d^{(i)}$ - estimated error for timestep i,
$\mathbf{y}_n^{(i)}$ - corrector solution for the $(n)^{th}$ iteration of timestep i, and
$\mathbf{y}_0^{(i)}$ - predictor solution for timestep i.

The above estimate assumes the iteration error is much smaller and therefore negligible compared to the truncation error of the difference formula. Selection of a new stepsize (Equation 4.4) is based on the local error estimate for the current timestep [38].

$$h' = 0.95h \left[\frac{\epsilon}{d^{(i)}} \right]^{\frac{1}{2}} \quad (4.4)$$

where h' - new stepsize,
h - current stepsize, and
ϵ - allowable step error.

If the current stepsize is less than or equal to the new stepsize (i.e., the local error criterion has been satisfied for the current timestep), then the new stepsize is used for the next timestep. Otherwise, the solution for the current timestep is recomputed using the new (i.e., smaller) stepsize. A 95% factor is used in the stepsize formula (Equation 4.4) to allow for any inaccuracies in the error estimate.

4.2 Iterative Methods

Discretization of the equations of motion produces a set of coupled non-linear algebraic equations. The only viable approaches for solving this set of algebraic equations are Newton's method and derivatives of Newton's method if the original differential equations are stiff [38]. Application of Newton's method requires formulation of the Jacobian for the system of algebraic equations. The linearized algebraic equations (obtained from applying Newton's method to the original non-linear equations) are solved using LU decomposition [50].

The Jacobian is dependent on foot-soil interactions and different combinations of soil loading conditions under each foot (e.g., vertical loading combined with lateral unloading) affect its formulation. Unloading characteristics (i.e., discontinuous slope of the force-deflection curve) and intermittent contact of the foot-soil modeling prohibit symbolic definition of the Jacobian. A difference approximation (Equation 4.5) is used to calculate the Jacobian where each coordinate is perturbed while maintaining all other dofs stationary.

$$\frac{\partial \mathbf{J}}{\partial y_j} = \frac{\mathbf{f}^* - \mathbf{f}}{\delta} \quad (4.5)$$

where $\frac{\partial \mathbf{J}}{\partial y_j}$ - j^{th} column of the Jacobian,
$\mathbf{f}$ - residual vector for trial solution,
$\mathbf{f}^*$ - residual vector for perturbed solution,
$\mathbf{y}^*$ - perturbed solution $= (y_1, y_2 \ldots y_j + \delta \ldots y_n)$, and
δ - perturbation of coordinate j.

The equations of motion characterizing legged locomotion on natural terrain are composed of inertial, joint and foot-soil interaction force contributions (see Equation 3.1). The calculation of inertial forces is computationally intensive relative to the other components and the observed change in inertial force contributions to the Jacobian is minimal between iterations (i.e., small solution variations). Foot-soil interaction forces exhibit the largest variation between solution iterations.

Based on this assessment, a full Newton solution (where all force contributions to the Jacobian are updated) is used for the first iteration of a timestep. The inertial force contribution from the first iteration is combined with updated joint and foot-soil contributions to formulate the Jacobian for subsequent iterations. This approach is defined here as the altered Newton solution (not to be confused with a modified Newton method) and no effect on the rate of solution convergence compared to the full Newton method has been detected.

Solution stability of the altered Newton solution is monitored by calculating the percentage difference between the current and previous convergence ratios[1]. When the difference in convergence ratios becomes less than some specified limit, the solution is considered to have stabilized (i.e., indicating a minimal change in the Jacobian) and the iterative solution switches from the altered Newton method to a chord solution. The Jacobian is not updated between iterations for the chord method and does not require decomposition (decomposed matrices are stored from an earlier altered Newton iteration) which further reduces iteration times. The iteration error for the chord solution [39] is calculated by the following equation.

$$e_{chord} = \frac{e_{calc}}{1 - ratio} \quad (4.6)$$

where e_{calc} - unmodified root mean square iteration error, and
$ratio$ - convergence ratio between successive iterations.

---

[1] Convergence ratio for the $(n)^{th}$ iteration of a timestep is calculated as the ratio of root mean square errors for the $(n)^{th}$ and $(n-1)^{th}$ iterations.

The convergence ratio for the chord solution is continually monitored and if this ratio exceeds an economical limit (defined by another specified parameter), then the iterative solution reverts to the altered Newton method.

4.3 Implementation

The application of boundary conditions (i.e., specified joint force histories or trajectories) to the equations of motion for a model of legged locomotion on natural terrain generally produces an uncoupled set of DAE's. Solution procedures are applied to the equations of motion in their original form (see Equation 3.1) without conversion to standard DAE form. Effectively, the algebraic joint force equations are solved concurrently with the differential equations that have been discretized into algebraic equations through substitution of backward difference expressions. The number and order of the equations is constant for a specific walker configuration (irrespective of the imposed boundary conditions) and re-ordering or transformations are not required for this approach.

The backward difference expression for the acceleration of a generalized coordinate (Equation 4.7) is obtained from application of the implicit Euler formula.

$$\ddot{\mathbf{y}}^{(i)} = \frac{\mathbf{y}^{(i)} - 2\mathbf{y}^{(i-1)} + \mathbf{y}^{(i-2)}}{h^2} \tag{4.7}$$

A trial solution exists (i.e., $\mathbf{y}^{(i)}$ is estimated) for each iteration so that velocities and accelerations of coordinates having non-specified trajectories can be calculated. Substitution of coordinate positions, velocities and accelerations (both calculated and specified), and applied joint forces into the equations of motion produces the residual vector for the trial solution. This vector defines the right hand side of the algebraic equations which are to be solved. The Jacobian is determined by differencing residual vectors for trial and perturbed solutions, respectively (see Section 4.2). Therefore, the whole solution process is summarized as numerically evaluating residual vectors using the same form of the equations of motion for all problem variations.

Columns of the Jacobian corresponding to coordinates having specified trajectories (i.e., joint force is unknown) are easily determined since $\partial \mathbf{J}/\partial F_j = 1$ (Equation 3.1). Specified perturbations are used to calculate the remaining columns of the Jacobian. Two parameters are defined for each of the two dof types (i.e., 6 body dofs and m joint dofs). The first parameter prescribes the relative perturbation for a coordinate with non-zero position (i.e., greater than .01) while the second parameter defines an absolute perturbation for a coordinate having a near-zero position. A relative perturbation applied to a zero position results in no change at all. The choice of parameters has a profound influence on the convergence ratio of the iterative solution as shown in Appendix B.

Mechanism states for three timesteps (i.e., previous, current and future states) required by the second order system equations must be specified to begin a simulation. The specified states must satisfy equilibrium conditions, otherwise the solution may not converge for the initial timestep. The indeterminacy of legged locomotion makes it nearly impossible to ascertain appropriate initial conditions for a mechanism resting on compliant terrain without prior calculations.

Acceptable mechanism states are calculated by a settling simulation where the mechanism is suspended slightly above the terrain, and then released to oscillate and settle to an equilibrium position. Initial mechanism states for a settling simulation are known since the mechanism is stationary with the current and previous states being the same. The future state is equated to the current mechanism state. A restart file is produced at the end of a settling simulation providing the required state information used for subsequent simulations. The future mechanism state in the restart file is predicted with the explicit Euler formula (Equation 4.2).

The initial stepsize specified for a settling simulation must be less than or equal to the stepsizes determined by the solution routine for the immediately following timesteps (thus satisfying the local error criterion). A few trial simulations quickly identify the appropriate initial stepsize for a settling simulation. Subsequent simulations use an initial stepsize provided in the appropriate restart file.

The stepsize is divided by a specified factor if more than ten iterations are required for a timestep; the stepsize will be reduced recursively up to four times in an attempt to converge to a solution for a given timestep. Impulsive loadings (i.e., near discontinuities), which require small stepsizes near the time of load application, are accommodated with this procedure. The solution routine has proven to be extremely robust and if convergence can not be obtained, then model parameters are probably incorrectly defined (e.g., discontinuous joint motions).

The multiplier by which a stepsize can be increased is limited by a specified factor because experience has shown that too large an increase may result in excessive iterations even though the local error criterion is satisfied. Similarly, the maximum stepsize may be specified if excessive iterations occur. Controlling the maximum stepsize and the proportion of stepsize increase eliminates the stepsize cycling (i.e., automatic reduction of a larger stepsize because of excessive iterations) that occurred in the initial applications of these solution procedures. Setting the stepsize increase and decrease factors equal to unity results in a constant stepsize which is useful for control system simulations. (A constant stepsize enables simulation of a constant sensor sampling rate.)

Chapter 5

Application to a Prototype Walking Machine

The modeling and solution procedures discussed in previous chapters are applied to the AMBLER prototype walking machine. The design and physical parameters of this mechanism configuration are described prior to their use. The dynamic equations for legged locomotion on natural terrain are specialized to take advantage of the AMBLER's orthogonal legged configuration for increased efficiency. A massless leg model is developed where the leg inertias are reallocated to the body resulting in a simplified formulation, useful for obtaining fast estimates of mechanism performance.

5.1 AMBLER Design

The AMBLER is an orthogonal legged, hexapod walking machine [47] with the configuration shown in Figure 5.1. The body of the walking machine consists of two posts whose upper ends are rigidly connected by a braced crossmember; three legs attach to each body post. Each leg has two members that move in a lateral plane and a vertical member that extends orthogonally from this plane (Figure 5.2). In outward order from a body post, the leg members are defined as inner, outer and vertical links, respectively.

The rotational shoulder joint connects a leg to the body post and the prismatic joint between lateral leg members is an elbow joint. The vertical link attaches to the outer member at the vertical prismatic joint and the ankle joint connects the foot to the leg. The unactuated ankle joint is free to rotate about the vertical axis but is prevented from rotating about the lateral axes (consistent with the discussion in Section 3.3.3). Both vertical and lateral prismatic joints slide on linear bearings. Bearing rails are mounted to the extending member while bearings are attached to the fixed member.

The shoulder, elbow and vertical joints are powered by brushless dc servo

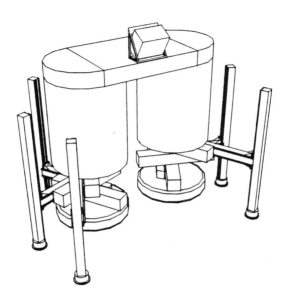

Figure 5.1: AMBLER Walking Machine Configuration

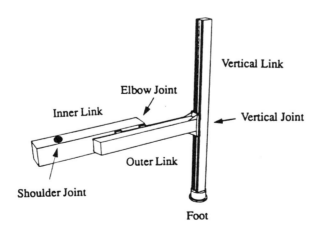

Figure 5.2: AMBLER Leg Design

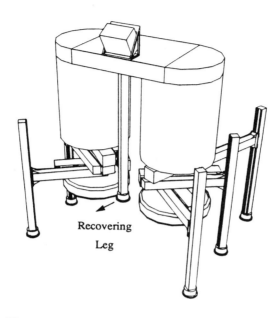

Figure 5.3: AMBLER Recovering Leg Motions

motors driving through torque multiplying gearboxes; the gearboxes and motor amplifiers are 80% and 90% efficient, respectively. Torque from the shoulder joint gearbox drives a spur gear for a total ratio of 424.7:1. The rotary output of the elbow and vertical joint gearboxes are converted to linear motion with a rack and pinion where the rack is affixed to the extending member. The elbow and vertical joint drivetrains both have a total ratio of 62.1:1.

Circulating crawl gaits are the intended mode of walking for the AMBLER, where only one leg is off the ground at a time ensuring maximum support stability should one or more footholds fail. Initial gaits intended for AMBLER are composed of discrete body move and leg recovery motions. During the body move portion of a gait cycle, the mechanism body is propelled in the lateral plane by combinations of shoulder and elbow joints; vertical joint positions are held fixed. When body progress has been completed, the foot of the recovering leg is lifted from the terrain and the leg recovers from the back of the stance to the front of the stance by passing between the body posts as shown in Figure 5.3. Once the leg has finished recovering, the foot is planted and the cycle is repeated.

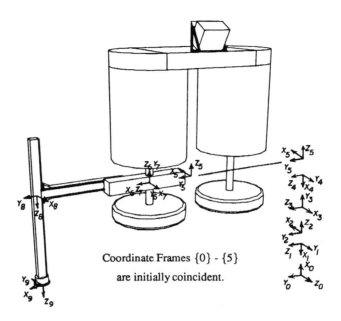

Figure 5.4: AMBLER Coordinate Frames

5.2 Geometric Representation

Approximations are made in the mechanism representation to simplify the kinematic equations and resulting dynamic model. All three of the legs which connect to a body post are considered to attach at the elevation of the middle leg and the offset between lateral leg members at the elbow joint is neglected. These offsets (both vertical and lateral at the shoulder and elbow joints, respectively) are judged to be small (i.e., 25.4 and 27.94 cm, respectively) relative to the scale of the machine (approximately 5 m) and, therefore, to have little effect on the mechanism's response. All of the legs are considered to be identical to each other because of these assumptions.

There are 6 body dofs plus 3×6 appendage joint dofs for a total of 24 generalized coordinates. The assignment of coordinate frames is shown in Figure 5.4 with the corresponding Denavit-Hartenberg parameters given in Table 5.1. The definition of the distance dl_i in coordinate frame $^5A_6$ varies according to the application of the transformation matrix as explained in the following paragraph.

The distance dl_i is equal to zero when calculating the body partial derivative matrices (U_{6j} and U_{6km} of Section 3.2) placing the T_6 coordinate frame coincident with the body frame. The distance dl_i is equal to $lb/2$ (half the distance between body posts) for calculation of the leg partial derivative matrices when the leg being considered is attached to the positive body post (i.e.,

Application to a Prototype Walking Machine

Link	θ	d	a	α
1	180°	δ_z	0	90°
2	−90°	δ_y	0	90°
3	−90°	δ_x	0	90°
4	$\theta_z - 90°$	0	0	90°
5	$\theta_y - 90°$	0	0	90°
6	$\theta_x + 90°$	0	dl_i	0
7	$\theta_{1_i} + 90°$	0	0	90°
8	0	d_{2_i}	0	90°
9	90°	d_{3_i}	0	0

Table 5.1: AMBLER Denavit-Hartenberg Parameters

starboard side). In this case, the $\mathbf{T}_6$ coordinate frame is coincident with the positive body post. The distance dl_i is equal to $-lb/2$ for a leg connected to the negative body post (i.e., port side).

The global position of a foot determines the soil deformation state so that foot-soil interaction forces can be calculated. Successive transformations (i.e., $\mathbf{T}_9$) produced the global positions of Foot i given below.

$$
\begin{aligned}
x_{foot_i} &= \delta_x - d_{3_i} \cos\theta_z \cos\theta_y + dl_i \cos\theta_z \sin\theta_y \sin\theta_x - dl_i \cos\theta_x \sin\theta_z \\
&\quad + d_{2_i} \cos\theta_z \cos\theta_x \sin\theta_y \sin\theta_{1_i} + d_{2_i} \cos\theta_z \cos\theta_{1_i} \sin\theta_y \sin\theta_x \\
&\quad - d_{2_i} \cos\theta_x \cos\theta_{1_i} \sin\theta_z + d_{2_i} \sin\theta_z \sin\theta_x \sin\theta_{1_i} \\
y_{foot_i} &= \delta_y - d_{3_i} \cos\theta_y \sin\theta_z + dl_i \cos\theta_z \cos\theta_x + dl_i \sin\theta_z \sin\theta_y \sin\theta_x \\
&\quad + d_{2_i} \cos\theta_z \cos\theta_x \cos\theta_{1_i} - d_{2_i} \cos\theta_z \sin\theta_x \sin\theta_{1_i} \qquad (5.1)\\
&\quad + d_{2_i} \cos\theta_x \sin\theta_z \sin\theta_y \sin\theta_{1_i} + d_{2_i} \cos\theta_{1_i} \sin\theta_z \sin\theta_y \sin\theta_x \\
z_{foot_i} &= \delta_z + d_{3_i} \sin\theta_y + dl_i \cos\theta_y \sin\theta_x + d_{2_i} \cos\theta_y \cos\theta_x \sin\theta_{1_i} \\
&\quad + d_{2_i} \cos\theta_y \cos\theta_{1_i} \sin\theta_x
\end{aligned}
$$

Foot forces are initially determined in the global frame consistent with the terrain surface definition. These global foot forces (both forces and moments) are rotated into the body coordinate frame for subsequent calculation of generalized forces created by foot-soil interactions. Inverting the product of the body transformations ($\mathbf{T}_6$) and a constant transformation to realign coordinate axes with the body frame ($^6\mathbf{A}_B$) produces the required rotation matrix. The constant transformation $^6\mathbf{A}_B$ is shown in Figure 5.5 and the resulting rotation matrix is given below.

A General Model of Legged Locomotion on Natural Terrain

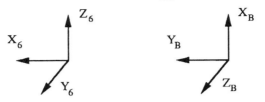

note: coordinate frames {6} & {B} are coincident

$$^6\mathbf{A}_B = \begin{bmatrix} 0 & 1 & 0 & 0 \\ 0 & 0 & 1 & 0 \\ 1 & 0 & 0 & 0 \\ 0 & 0 & 0 & 1 \end{bmatrix}$$

Figure 5.5: Constant Transformation to Realign Coordinate Axes

$$^B\mathbf{R}_0 = \qquad\qquad\qquad\qquad\qquad\qquad\qquad\qquad (5.2)$$

$$\begin{bmatrix} \cos\theta_y \cos\theta_z & \cos\theta_y \sin\theta_z & -\sin\theta_y \\ \sin\theta_x \sin\theta_y \cos\theta_z - \cos\theta_x \sin\theta_z & \cos\theta_x \cos\theta_z + \sin\theta_x \sin\theta_y \sin\theta_z & \sin\theta_x \cos\theta_y \\ \cos\theta_x \sin\theta_y \cos\theta_z + \sin\theta_x \sin\theta_z & -\sin\theta_x \cos\theta_z + \cos\theta_x \sin\theta_y \sin\theta_z & \cos\theta_x \cos\theta_y \end{bmatrix}$$

The generalized body forces shown below, which result from foot-soil interactions, are calculated from the transformed foot forces.

$$\begin{aligned}
F_x &= \sum_{i=1}^{6} f_{x_i} \\
F_y &= \sum_{i=1}^{6} f_{y_i} \\
F_z &= \sum_{i=1}^{6} f_{z_i} \\
M_x &= \sum_{i=1}^{6} [m_{x_i} - f_{y_i} d_{2_i} \sin\theta_{1_i} + f_{z_i}(d_{2_i} \cos\theta_{1_i} + dl_i)] \\
M_y &= \sum_{i=1}^{6} [m_{y_i} + f_{x_i} d_{2_i} \sin\theta_{1_i} + f_{z_i} d_{3_i}] \\
M_z &= \sum_{i=1}^{6} [m_{z_i} - f_{x_i}(d_{2_i} \cos\theta_{1_i} + dl_i) - f_{y_i} d_{3_i}]
\end{aligned} \qquad (5.3)$$

where $f_{x_i}, f_{y_i}, f_{z_i}, m_{x_i}, m_{y_i}, m_{z_i}$ - foot forces and moments on Leg i that have been rotated into the body coordinate frame, and
$F_x, F_y, F_z, M_x, M_y, M_z$ - generalized body forces.

Application to a Prototype Walking Machine

The above expressions are general in that 6 components of foot forces (i.e., 3 forces plus 3 moments) are considered even though rotational foot-soil interactions are not currently represented. Generalized joint forces for Leg i produced by foot-soil interactions are determined below.

$$\begin{aligned}
F_{shoulder_i} &= f_{z_i} d_{2_i} \cos\theta_{1_i} - f_{y_i} d_{2_i} \sin\theta_{1_i} + m_{x_i} \\
F_{elbow_i} &= f_{y_i} \cos\theta_{1_i} + f_{z_i} \sin\theta_{1_i} \\
F_{vertical_i} &= -f_{x_i}
\end{aligned} \quad (5.4)$$

where $F_{shoulder_i}, F_{elbow_i}, F_{vertical_i}$ - generalized joint forces for Leg i.

5.3 Physical Parameters

The pseudo-inertia matrices [31] and center of mass vectors ($\mathbf{J}_i$ and $\bar{\mathbf{r}}_i$, respectively, in Equation 3.1) used to represent the body and leg links in the dynamic formulation are given in Table 5.2. These parameters are defined according to the coordinate frame convention discussed in Section 5.2. Joint damping and backdriving effects are considered to be dependent on the joint bending moment. Parameters used to model these phenomena (see Equations 3.2 and 3.3) are shown in Table 5.3.

Frictional effects at the ankle joint cannot be directly represented since a generalized coordinate is not defined for this location. Instead, ankle friction is modeled as a pseudo-rotational foot-soil interaction about the vertical leg axis. This twisting moment is already specified in the body coordinate frame so the transformation required of actual foot-soil interaction forces (Equation 5.2) does not apply to this component.

Flexibilities of a typically configured leg (i.e., elbow and vertical joints half extended) were calculated using standard structural analysis techniques. All joints were considered to be rigid and fixed boundary conditions were enforced at the shoulder joint simulating a rigid body structure. The aluminum leg members were modeled using rectangular beam elements with a uniform .3175 cm wall thickness which neglects cutouts, gussets, and the .635 cm thick walls that support all bearing rails. Inner and outer links were connected with rigid offsets of 40.64 cm located at each of the two bearing pads. Vertical and lateral loads were separately imposed on the leg structure and compliances in the direction of load application were calculated with the results shown in Figure 5.6. The vertical compliance of .0629 cm/kN and lateral compliance of .3689 cm/kN (larger of the two lateral compliances) are used to augment foot-soil representations.

5.4 Specialization of Dynamic Equations

The partial derivative matrices ($\mathbf{U}_{ij}$ and $\mathbf{U}_{ikm}$) were symbolically evaluated using the coordinate frame descriptions and associated transformation matrices

$$\mathbf{J}_6 = \begin{bmatrix} 308.175 & 0 & 0 & 0 \\ 0 & 622.035 & 0 & 0 \\ 0 & 0 & 825.515 & 1025.41 \\ 0 & 0 & 1025.41 & 907.2 \end{bmatrix} \quad \bar{\mathbf{r}}_6 = \begin{bmatrix} 0 \\ 0 \\ 1.1303 \\ 1 \end{bmatrix}$$

$$\mathbf{J}_7 = \begin{bmatrix} .26 & 0 & 0 & 0 \\ 0 & .26 & 0 & 0 \\ 0 & 0 & 12.55 & 6.6 \\ 0 & 0 & 6.6 & 40. \end{bmatrix} \quad \bar{\mathbf{r}}_7 = \begin{bmatrix} 0 \\ 0 \\ .1651 \\ 1 \end{bmatrix}$$

$$\mathbf{J}_8 = \begin{bmatrix} .04 & 0 & 0 & 0 \\ 0 & 20.8 & 0 & -18.034 \\ 0 & 0 & .07 & 0 \\ 0 & -18.034 & 0 & 20. \end{bmatrix} \quad \bar{\mathbf{r}}_8 = \begin{bmatrix} 0 \\ -.9017 \\ 0 \\ 1 \end{bmatrix}$$

$$\mathbf{J}_9 = \begin{bmatrix} .125 & 0 & 0 & 0 \\ 0 & .125 & 0 & 0 \\ 0 & 0 & 160.735 & -76.24 \\ 0 & 0 & -76.24 & 58.97 \end{bmatrix} \quad \bar{\mathbf{r}}_9 = \begin{bmatrix} 0 \\ 0 \\ -1.2929 \\ 1 \end{bmatrix}$$

$$\mathbf{J}_i = \begin{bmatrix} kg\,m^2 & kg\,m^2 & kg\,m^2 & kg\,m \\ kg\,m^2 & kg\,m^2 & kg\,m^2 & kg\,m \\ kg\,m^2 & kg\,m^2 & kg\,m^2 & kg\,m \\ kg\,m & kg\,m & kg\,m & m \end{bmatrix} \quad \bar{\mathbf{r}}_i = \begin{bmatrix} m \\ m \\ m \\ 1 \end{bmatrix}$$

All pseudo inertia terms are calculated to three significant figures while all cg locations are calculated to four figures.

Table 5.2: AMBLER Physical Properties

Application to a Prototype Walking Machine

Joint	Coulomb Friction Value F_{cf} (N or Nm)	Asymptotic Backdriving Force F_{ab} (N or Nm)
shoulder	$15 + |M|/625.7322$	$26 + |M|/312.8661$
elbow	$44.48 + |M|/45.7201$	$44.48 + |M|/45.7201$
vertical	$22.24 + |M|/45.7201$	$22.24 + |M|/45.7201$

M - bending moment acting on joint (Nm)

Table 5.3: AMBLER Joint Damping and Backdriving Parameters

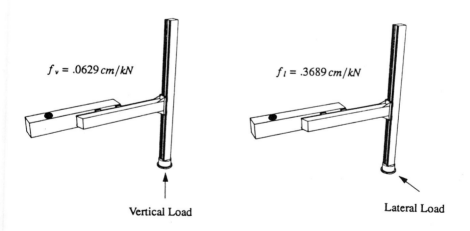

Figure 5.6: AMBLER Structural Flexibilities

$$\mathbf{U}_{i,1} = \begin{bmatrix} 0 & 0 & 0 & 0 \\ 0 & 0 & 0 & 0 \\ 0 & 0 & 0 & 1 \\ 0 & 0 & 0 & 0 \end{bmatrix}$$

$$\mathbf{U}_{i,2} = \begin{bmatrix} 0 & 0 & 0 & 0 \\ 0 & 0 & 0 & 1 \\ 0 & 0 & 0 & 0 \\ 0 & 0 & 0 & 0 \end{bmatrix} \quad where\ i = 6\ldots 9$$

$$\mathbf{U}_{i,3} = \begin{bmatrix} 0 & 0 & 0 & 1 \\ 0 & 0 & 0 & 0 \\ 0 & 0 & 0 & 0 \\ 0 & 0 & 0 & 0 \end{bmatrix}$$

$$\mathbf{U}_{ikm} = 0 \quad where\ i = 6\ldots 9,\ k = 1\ldots 3,\ m = 6\ldots i$$

$$\mathbf{U}_{9k9} = 0 \quad where\ k = 6\ldots 9$$

$$\mathbf{U}_{i88} = 0 \quad where\ i = 8, 9$$

Table 5.4: Degenerate Partial Derivative Matrices

described in Section 5.2 to determine which matrices reduce to simple expressions. Approximately 60% of the matrices (90 out of 156) are constant or identically zero; the degenerate matrices are identified in Table 5.4. The inertial terms in the dynamic equations which are products of the zero partial derivative matrices are not computed (since they are also zero) and the constant matrices are explicitly assigned, reducing computational effort.

5.5 Massless Leg Model

A massless leg model is formulated by reallocating leg inertias to the body and updated body parameters are calculated for the first iteration of each timestep. (Modified body properties are not calculated for subsequent iterations because small variations in leg positions have little effect on the approximation.) Only the body dynamic contributions to the system equations are calculated for this model (leg dynamic contributions are zero since the legs are massless) which combined with the specialization of the dynamic equations results in almost two orders of magnitude reduction in computational effort (see Section 6.6).

The parallel axis theorem [40] is not applicable for reallocating leg inertias because shoulder joint rotations skew the link axes relative to the body axes.

Application to a Prototype Walking Machine

Leg contributions to the body inertial parameters are calculated by specifying leg member positions in the body coordinate frame and substituting these expressions into the body inertia equations for subsequent integration. Transformation of link inertial parameters for the AMBLER configuration is described in Appendix C.

The massless leg model is appropriate for simulations where all leg joints have specified trajectories. This model becomes singular for simulations where forces (e.g., joint and foot-soil interaction) are applied to a leg and the resulting leg motions are calculated (not specified) because a massless leg would experience an infinite acceleration under these conditions. The relative accuracy of the massless leg model is discussed in Chapter 6.

Chapter 6

Verification Studies

The modeling and solution procedures discussed in previous chapters are verified by simulating physical cases that isolate features of the legged locomotion model for confirmation of principles such as symmetry, equilibrium, and other measures of physical reasonableness. The AMBLER mechanism is configured in a symmetric posture for all verification cases, which results in the expectation that a symmetric solution should be calculated with sufficient accuracy.

Foot forces are sensitive to solution variations because of the relatively large stiffnesses involved in the modeling of foot-soil interactions. Therefore, any deviations from a symmetric solution are magnified in the foot force results. Discrepancies in the symmetry of calculated foot forces are less than .01% for all verification cases, which is excellent solution accuracy. It was determined by initial verification runs that maintaining sixteen significant digits during the calculations is required to produce this degree of symmetric results.

The first set of verification studies are flat settlement examples where the mechanism is released from a position slightly above the ground to oscillate and settle to an equilibrium position on the soil. These simulations isolate the vertical foot-soil interactions because appreciable lateral effects are not developed. Next, settlement on a sloped surface is considered which induces significant lateral foot forces and displacements without the complexities of actuated joint motions. Body move examples (beginning with settled mechanism conditions) are investigated because joint motions couple with vertical and lateral foot-soil interactions to propel the body. Finally, leg recovery motions are simulated to determine model performance for foot liftoff and placement conditions.

The verification studies exercise all mechanism motions so that subsequent model applications (e.g., gait cycle simulations) are merely combinations of previously verified actions. Model and solution parameters used for the verification analyses are presented below; calculated results are discussed for each simulation. The final section of this chapter discusses the computational performance of the legged locomotion models by observing computation times required by the full Newton, altered Newton, and chord iterative solution methods.

The legged locomotion model produces results that are physically reasonable for all cases considered and equilibrium conditions are satisfied for the steady-state foot forces. The massless leg model accurately predicts mechanism responses for the flat and sloped settlement examples verifying the formulation of the simplified model. Body propulsion results are overestimated by the massless leg model because leg members experience minimal motions (requiring little power) yet their inertias are reallocated to the body; a heavier equivalent body must then be propelled. The massless leg model is considered sufficiently accurate for use in preliminary investigations. Based on the verification results of this chapter, it is concluded that the model of legged locomotion on natural terrain and associated solution procedures developed in this monograph are verified and suitable for general application.

6.1 Model and Solution Parameters

The AMBLER configuration discussed in Chapter 5 is modeled for the verification analyses. Coefficients for the joint damping and backdriving representations (c_d and c_b, in Equations 3.2 and 3.3, respectively) are both assigned a value of $300\,sec^{-1}$ which provides a quick transition from zero force to the steady-state values representative of stiction. This transition modeling produces 99.5% of the steady state response for rotational and prismatic joint velocities of $1\,deg/sec$ and $1.8\,cm/sec$, respectively. The gravitational constant for Earth gravity is specified as $9.81\,m/sec^2$.

The modeled soil is a homogeneous, compact sand with the properties given in [45]. Structural deflections are added to soil sinkages, and a least squares fit of the bilinear equation produces the vertical force-deflection curve and associated parameters shown in Figure 6.1. Similarly, lateral soil displacements subject to a $2224\,N$ vertical load (which is considered typical) are supplemented with additional structural deflections and a least squares fit of the exponential transition to Coulomb's equation results in the curve and transition parameter shown in Figure 6.1. The original shear parameters were measured using a bevameter [24] which is a plate annulus having radial grousers that cause the soil (not the plate-soil interface) to fail during loading. Applying this data for modeling foot-soil interactions implicitly assumes an adequate sole design that eliminates slippage at the foot-soil interface.

The limit on local step error (i.e., ϵ in Equation 4.3) used in all analyses is 1×10^{-5} and the maximum iteration error is specified as 1×10^{-8}. The iteration error is negligible compared to the step error as assumed in the solution formulation of Section 4.1. The set of perturbation parameters identified in Appendix B is used for calculating the Jacobian. The maximum stepsize increase factor is conservatively set equal to 1.25 which is less than a value of 1.38 that produced stepsize cycling in initial simulations. Similarly, a 1.25 stepsize reduction factor is used, which has proven acceptable in accommodating impulsive loadings.

Verification Studies

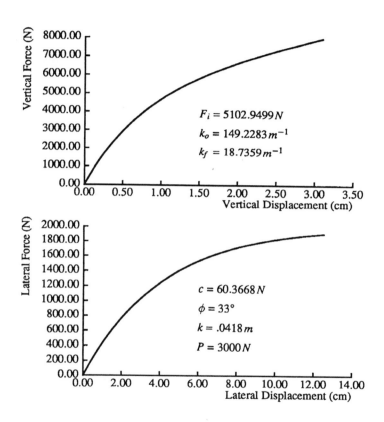

Figure 6.1: Force-Deflection Relationships for a Compact Sand

The iterative solution switches from the altered Newton method to the chord method when the difference in successive convergence ratios becomes less than 5%. An efficiency factor of 5 is used to determine the maximum convergence ratio for the chord method that is economical relative to the altered Newton approach. If this maximum value is exceeded, then the iterative solution reverts to the altered Newton method.

Three different mechanism driving configurations are considered for the verification studies:

Case 1 - full model (i.e., not massless leg model) with all shoulder and elbow joints powered

Case 2 - massless leg model with all shoulder and elbow joints powered

Case 3 - full model with the shoulder and elbow joints of Legs 2 and 5 powered (all other shoulder and elbow joints are backdriven)

The vertical leg joints remain fixed for all cases. Because of symmetry in the mechanism stance and calculated solutions, results are provided only for legs connecting to the starboard body post (i.e., Legs 1, 2, and 3).

6.2 Flat Settlement Examples

The stance used for simulating mechanism settlement on a flat soil surface is shown in Figure 6.2, with all vertical leg lengths fixed at $1.0914\ m$. Simulations begin with the mechanism in suspension slightly above the ground and then released to oscillate until an equilibrium position is reached. A $.001\ sec$ initial stepsize is used for these simulations which satisfies the local step error for the first timestep. Calculated results are given in Figures 6.3, 6.4, and 6.5 for Cases 1, 2, and 3, respectively.

In all cases, the mechanism initially accelerates downward at a rate equal to gravity, then oscillates with decaying amplitude to a steady-state position. The non-conservative foot-soil modeling produces the decaying response by dissipating energy from the system. Mechanism rocking occurs because the uneven fore-aft weight distribution generates lateral foot forces and uneven vertical foot forces that combine to rotate the mechanism. The maximum body pitch accelerations occur simultaneously with the development of both the maximum vertical and lateral foot forces. However, the clockwise effect of the lateral forces somewhat offsets the counterclockwise action of the vertical forces. The final body pitch rotation of $.0856°$ for Cases 1 and 2 (full and massless leg models with all joints locked), and $.0934°$ for Case 3 (full model with Legs 2 and 5 fixed) are reasonable considering that the body position is located towards the rear of the stance, biasing the mechanism weight distribution in this direction. The unpowered joints of Case 3 prevent the development of traction forces in corresponding legs that resist body rotations, resulting in the larger final pitch rotation.

Verification Studies

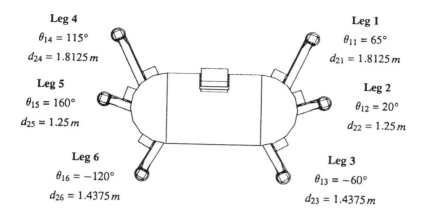

Figure 6.2: Mechanism Stance Used for Flat Settlement Examples

In Cases 1 and 2, all joints are fixed and the pitch body rotations create lateral soil deformations and forces that are proportional to the distance of a foot from the rotation axis. Lateral foot forces for Case 3 (full model with Legs 2 and 5 fixed) do not exhibit this consistent response because released elbow and shoulder joints permit the corresponding legs to move laterally instead of deforming the soil. Therefore, the mechanism response is not as simple for this case. The maximum lateral foot forces for Case 3 are less than the corresponding forces for the other cases because the released joints reduce lateral soil deformations.

The summation of steady-state vertical foot forces near the end of each simulation is equal to the mechanism weight although minor oscillations are observed in the response. Since there is no structural damping in the legged locomotion model and the soil model is linear elastic after all non-linear deformations have occurred (i.e., near steady-state), the mechanism would oscillate indefinitely with no energy being dissipated from the system. Numerical damping inevitably introduced by the solution procedures [51] ultimately attenuates the response.

Calculated results for massless leg model of Case 2 are essentially identical (i.e., .0002 % accuracy) to the corresponding full model results of Case 1. This comparison verifies the massless leg model formulation because an implicit assumption, when reallocating leg inertias, is that the mechanism responds as a single rigid body. For the flat settlement simulations of Cases 1 and 2, all leg joints are fixed and the mechanism does respond as a single rigid body.

The decay times for all Case 3 mechanism responses are greater than the other cases because four legs are free to move, thus reducing lateral soil deformations. The energy dissipated by backdriving joints through small motions is less than the energy attributed to lateral soil deformations that would occur if

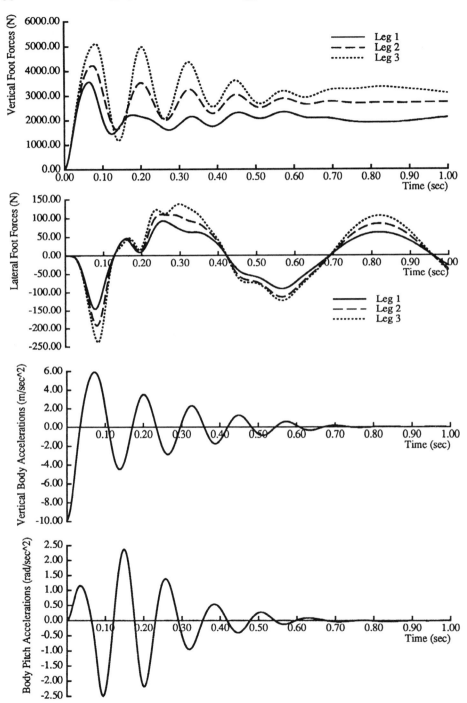

Figure 6.3: Flat Settlement Results for Case 1

Verification Studies 57

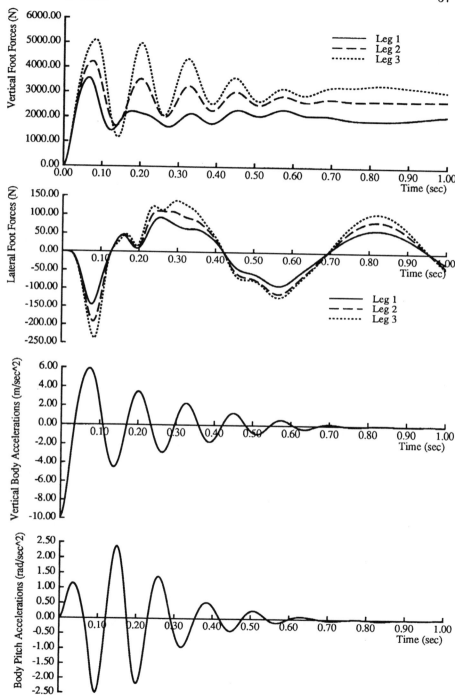

Figure 6.4: Flat Settlement Results for Case 2

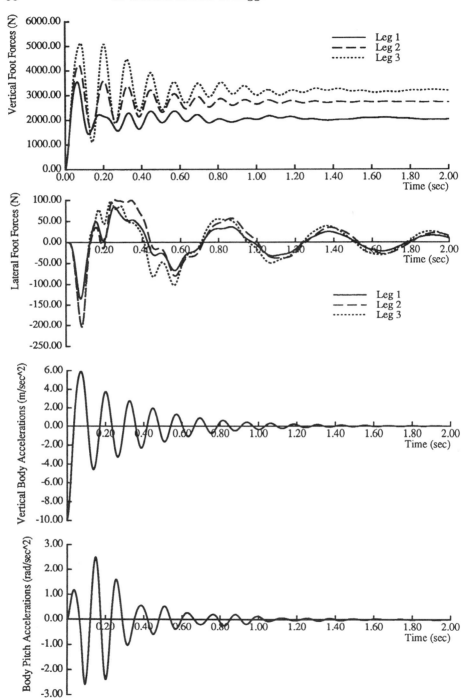

Figure 6.5: Flat Settlement Results for Case 3

Verification Studies

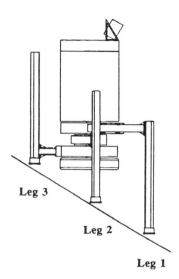

Figure 6.6: Mechanism Posture Used for Sloped Settlement Examples

all joints were kept fixed. Therefore, a longer decay time is observed for Case 3 because less energy is dissipated for each oscillation.

6.3 Sloped Settlement Examples

The same mechanism stance used for the flat settlement examples (Figure 6.2) is used for the sloped settlement examples. A 30° slope is considered where the mechanism axis of symmetry is parallel to the slope contour. The mechanism's posture on the sloped terrain is shown in Figure 6.6. Vertical lengths of $1.9127\,m$, $1.3052\,m$ and $.4689\,m$ for Legs 1, 2 and 3, respectively, conform to the ground slope so that the body is initially level when the feet contact the soil. An initial stepsize of $.001\,sec$, which satisfies the local step error for the first timestep, is used in the simulations. Calculated results are given in Figures 6.7, 6.8, and 6.9 for Cases 1, 2, and 3, respectively.

In Cases 1 and 2 (full and massless leg models with all joints locked), the mechanism initially accelerates downward, then oscillates with decreasing amplitude similar to the flat settlement results. Lateral accelerations are experienced during sloped settlement because sliding of the mechanism down the slope occurs as a combination of vertical and lateral motions. The lateral body acceleration decays rapidly for Cases 1 and 2, indicating a relatively small translation of the mechanism down the slope. The lateral foot forces for these cases peak at approximately $.11\,sec$, then drop to a lower level and oscillate to steady-state values. The mechanism experiences an initial surge down the slope followed by oscillations about the displaced position. Foot 1 lifts off the ground during the

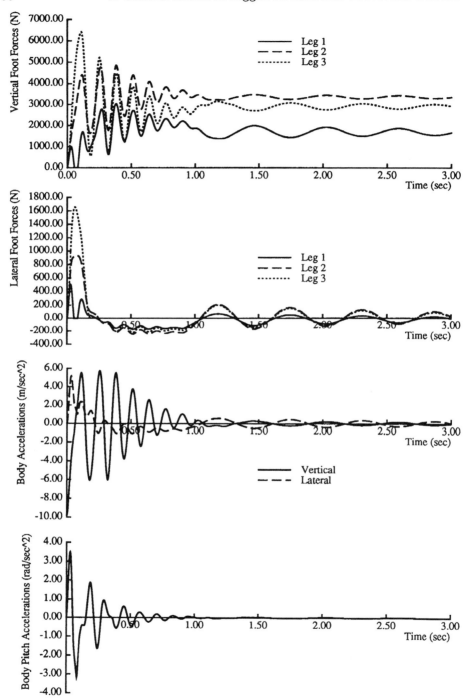

Figure 6.7: Sloped Settlement Results for Case 1

Verification Studies

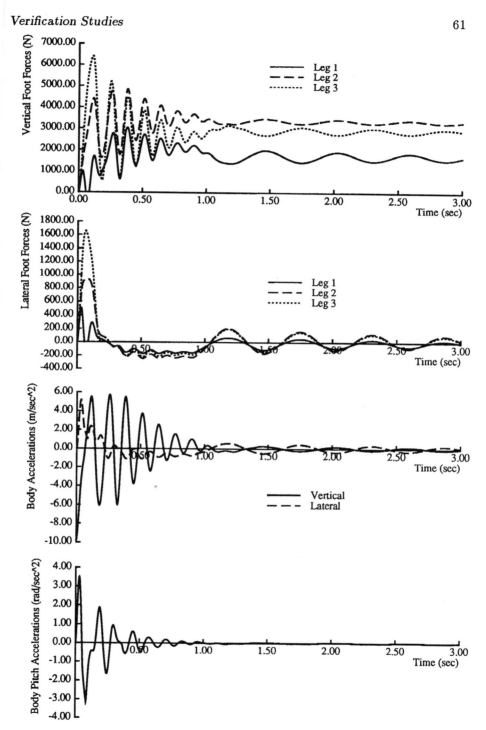

Figure 6.8: Sloped Settlement Results for Case 2

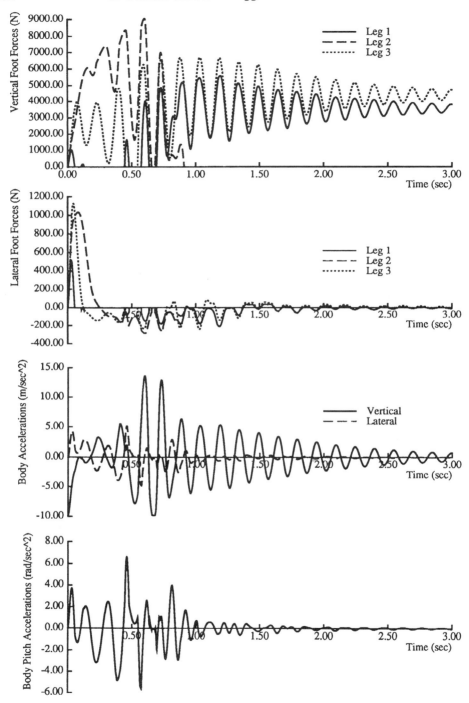

Figure 6.9: Sloped Settlement Results for Case 3

Verification Studies

initial surge for Cases 1 and 2 which causes impulsive loadings observed in the body pitch accelerations when the foot recontacts the ground.

The mechanism response for Case 3 (full model with Legs 2 and 5 fixed) is more complicated because unpowered joints permit lateral leg motions and increased translation of the mechanism down the slope. The vertical body acceleration is equal to gravity at the time of release, and diminishes as the soil deforms, but does not smoothly decay because of broken and re-established foot contacts. Feet 1 and 3 impact the ground (after liftoff) at approximately .6 sec creating foot forces that accelerate the mechanism upward at an acceleration of 1.5 g. This upward acceleration causes the mechanism to break all ground contacts at .65 sec and the mechanism is airborne for a short period of time. The mechanism subsequently touches down and Foot 2 (i.e., the "fixed" leg) never re-establishes contact because the released legs enable a lateral motion of the body away from the slope.

In Case 3 (full model with Legs 2 and 5 fixed), the mechanism hops down the slope producing pulsations that are observed in the body pitch accelerations. These high-frequency oscillations, arising from impulsive foot loadings, are superimposed upon a lower frequency rigid body response, thus characterizing the equations of motion as a set of stiff differential equations. The decay times of all Case 3 mechanism responses are greater than the other cases because legs are able to move laterally instead of deforming the soil.

The summation of vertical foot forces near the end of each simulation oscillates around the mechanism weight. Final slope translations (i.e., projection of combined vertical and lateral displacements onto the 30° ground surface) of the mechanism are .2393 m for Cases 1 and 2 (full and massless leg models with all joints locked), and .3044 m for Case 3 (full model with Legs 2 and 5 fixed). Once again, results calculated with the massless leg model in Case 2 are nearly identical to the corresponding full model results of Case 1. The sloped settlement results are not particularly credible because of the approximate modeling of sloped foot-soil interactions as discussed in Section 3.3.4. Instead, these examples demonstrate the robustness of the solution procedures and the appropriateness of the legged locomotion model (especially the set of generalized coordinates) for simulating tipover, freefall and foot contact/liftoff.

6.4 Body Move Examples

Final mechanism states from the flat settlement examples are used as initial conditions for the body move simulations. The desired motion is a .7 m translation of the body along the axis of symmetry for the mechanism stance (see Figure 6.2) over a 2 sec time interval. The quintic profile beginning and ending at zero velocity and acceleration, as shown in Figure 6.10, is specified for the body trajectory (i.e., position, velocity and acceleration history). Inverse kinematic equations of the lateral leg members are used to determine corresponding trajectories for the powered elbow and shoulder joints. The vertical

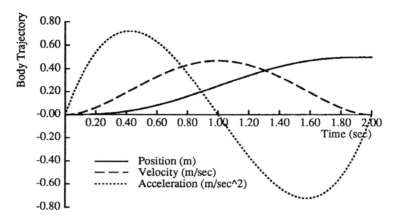

Figure 6.10: Desired Trajectory for Body Move Simulations

leg lengths remain fixed at $1.0914\,m$. The motions of powered joints are applied as boundary conditions to the model for subsequent calculation of the resulting body and unpowered leg motions. Foot slippage precludes exact reproduction of desired body trajectories. Calculated results are shown in Figures 6.11, 6.12 and 6.13 for Cases 1, 2 and 3, respectively.

In all cases, significant traction forces are developed at the feet of the powered legs (i.e., legs having actuated shoulder and elbow joints) that propel the body. The traction forces are positive, in general, during the acceleration phase $(0-1\,sec)$ and negative during deceleration $(1-2\,sec)$. The lateral foot forces for unpowered Legs 1 and 3 of Case 3 (full model with lateral joints powered for Legs 2 and 5) oppose the body motions which indicates backdriving of the unactuated shoulder and elbow joints for these legs. Maximum lateral foot forces calculated in Case 3 are approximately 160% greater than the results for the other cases because there are only two legs propelling the body and backdriving the unpowered joints.

An overturning moment is created by the lateral foot forces acting well below the mechanism's center of gravity. During the acceleration phase, positive traction forces cause a clockwise rotation of the mechanism that shifts the mechanism weight distribution from the forward legs to the rearward legs. Conversely, a clockwise rotation occurs during deceleration that loads the forward legs. This phenomenon of changing weight distribution caused by mechanism rocking is substantiated by the vertical foot force results.

Changes in vertical foot forces from initial to final mechanism positions for each simulation indicate the translation of the mechanism's center of gravity produced by the body movement. Since the body moves forward in the stance, it is expected that Leg 3 should unload and Leg 1 vertical forces will increase, which is observed in the results. Once again, summation of near steady-state vertical foot forces oscillates around the mechanism weight.

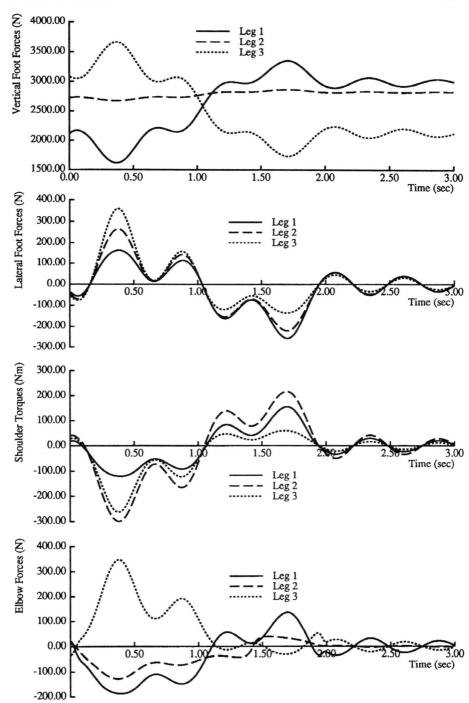

Figure 6.11: Body Move Results for Case 1

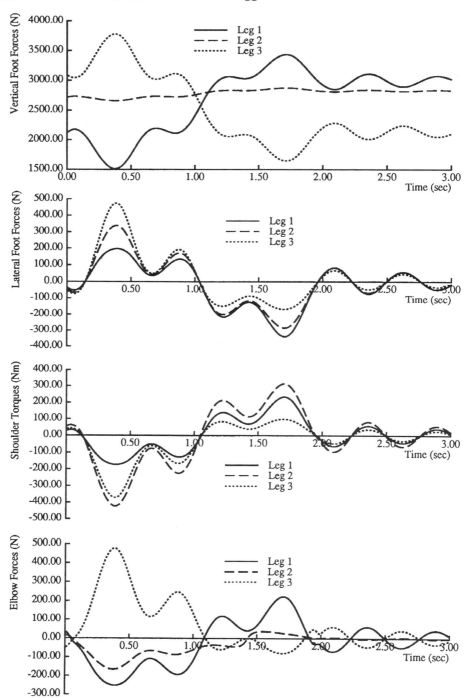

Figure 6.12: Body Move Results for Case 2

Verification Studies

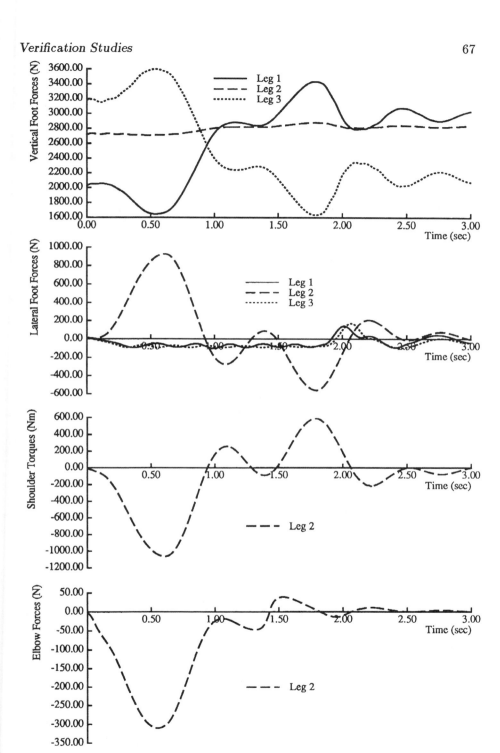

Figure 6.13: Body Move Results for Case 3

Time Interval	Corresponding Leg Motions
$0 - 1\ sec$	raise vertical members $5.08\ cm$
$1 - 2\ sec$	withdrawal elbow joints $6.25\ cm$
$2 - 4\ sec$	rotate shoulder joints $20°$ away from axis of symmetry
$4 - 5\ sec$	lower vertical members $3.1\ cm$

Table 6.1: Recovering Leg Motions

Vertical foot forces calculated with the massless leg model in Case 2 are within 3% of the results for Case 1 (full model with all lateral joints powered). This relatively good comparison indicates the mechanism's weight distribution is accurately represented with the simplified model. Lateral foot forces, shoulder torques, and elbow forces for Case 2 exceed the corresponding results of Case 1 by approximately 32%. This difference is attributed to the leg masses being reallocated to the body and subsequent propelling of the larger body mass. Some leg members do not translate appreciably during a body move (e.g., outer and vertical links), yet their inertial properties are lumped with those of the body. If leg masses were not reallocated to the body, the mechanism weight distribution would be underestimated because these members move with the body during rocking and vertical motions. The massless leg model predicts the mechanism weight distribution and, therefore, vertical soil deformations with high integrity at the cost of conservatively estimated propulsion effects.

6.5 Leg Recovery Example

The final mechanism state from the Case 1 flat settlement example (full model with all shoulder and elbow joints powered) is used as the initial state for the leg recovery study. The massless leg model (i.e., Case 2) is not used because the model becomes singular in this application. Case 3, which is the full model with powered shoulder and elbow joints for Legs 1 and 4, cannot recover the powered legs and is not considered. The motions for recovering Legs 1 and 4 of the Case 1 model are given in Table 6.1. Joint trajectories in each interval are specified according to a quintic profile beginning and ending at zero velocity and acceleration (see Figure 6.10). Calculated results for the leg recovery example are given in Figure 6.14.

The sudden withdrawal of Legs 1 and 4 cause a redistribution of the mechanism weight among the remaining legs which still contact the ground. Leg 2 is closest to the lifted leg and assumes a significant increase in vertical support forces while Leg 3 unloads due to a counterclockwise pitch of the mechanism. The vertical force redistribution from the broken ground contacts produces rocking oscillations of the mechanism that are observed in all results. Foot placement affects the vertical foot forces in a similar manner.

Verification Studies

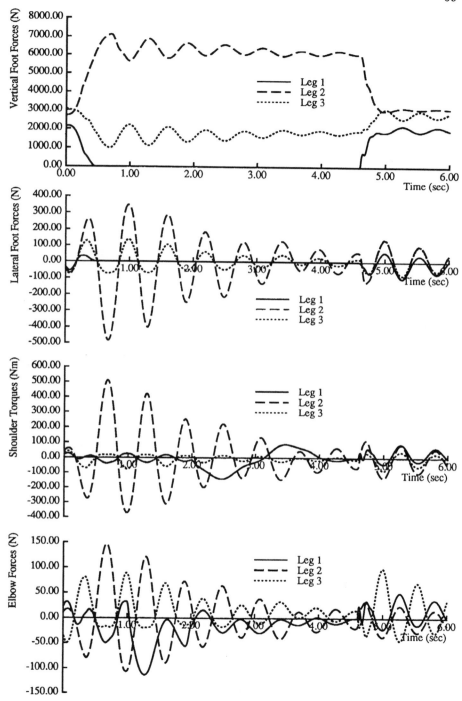

Figure 6.14: Leg Recovery Simulation Results

Summation of vertical foot forces near the end of the simulation fluctuates around the mechanism weight. Shoulder torques and elbow forces for the recovering leg exhibit oscillations during freeflight (i.e., zero foot forces) which are caused by rocking of the mechanism. This coupling is observed because foot-soil interactions were represented in a full dynamic formulation of the legged locomotion model. Even though the rates of foot withdrawal and placement are extreme for an actuator, the resulting impulsive loadings are not as severe as those produced by accidental foot liftoff/placement during the sloped settlement example.

6.6 Computational Performance

Although the model of legged locomotion developed in this monograph is not intended for real-time applications, computational performance is important because a more efficient model (at the same level of accuracy) can better support a research effort. The number of solution iterations required for each timestep varies according to many factors (e.g., stepsize, impulsive loading) so that timestep solution times are not an equitable measure of computational performance. Alternately, observing the run times required for the different iterative solution methods provides insight into the solution performance of the legged locomotion model.

Evaluation of the dynamic equations is the most expensive computation involved for each iterative solution method. A full Newton iteration (used for the first iteration) requires evaluation of the dynamic equations once to calculate the residual vector plus l evaluations to formulate the Jacobian matrix where l is the number of generalized coordinates having specified force histories (producing l differential equations which are to be solved). The altered Newton and chord iterations require evaluation of the dynamic equations only once to calculate the residual vector.

The computational effort required for each iteration method is proportional to the number of times the dynamic equations are evaluated. Models having all joints powered (i.e., Cases 1 and 2) have 6 differential equations corresponding to the body dofs and require 7 evaluations. The model of Case 3 has 8 unpowered joints with 14 differential equations necessitating 15 evaluations of the dynamic equations. Cycle times on a SUN 4/260[1] used for calculation of a full Newton iteration, altered Newton iteration, and a chord iteration for Cases 1, 2 and 3 are given in Table 6.2. Model and solution algorithms were programmed in Fortran 77 using double precision representations for all critical variables. As noted in the table, run times can be halved if the coriolis/centrifugal dynamic terms are not calculated because nearly half of the non-zero partial derivative matrices (see Section 5.4) are associated with these terms.

For all models, the chord solution is marginally faster than the altered Newton method for the same problem because joint and foot force contributions to

---

[1] SUN Microsystems SPARC-based workstation operating at 16.67 MHz

Verification Studies

	Computer Run Times[1],[2] (sec)		
Model	Iterative Solution Method		
Description	Full Newton	Altered Newton	Chord
Full Model - 6 DE's[3]	14.51	2.14	2.06
Massless Leg Model - 6 DE's[4]	.39	.13	.05
Full Model - 14 DE's[5]	31.06	2.22	2.08

(1) All timing studies were executed on a SUN 4/260 workstation.
(2) Above run times are halved if coriolis/centrifugal terms are not calculated.
(3) Case 1 model
(4) Case 2 model
(5) Case 3 model

Table 6.2: Iterative Solution Timing Studies

the Jacobian are not updated and the Jacobian does not require decomposition (decomposed matrices are stored from an earlier altered Newton iteration). The full Newton solution times are greater than the altered Newton/chord solution times for the same model by a factor that is approximately equal to the ratio of dynamic evaluations. Similarly, the ratio of full Newton run times for Case 3 relative to Case 1 is roughly equal to the ratio of evaluations for the corresponding models (i.e., 15:7). Iterative solution times for the massless leg model (Case 2) are almost two orders of magnitude less than the corresponding solution times for the full model (Case 1).

Chapter 7

Model Applications

This chapter describes applications of the legged locomotion model on natural terrain that demonstrate the utility and generality of the modeling and solution procedures developed earlier, and provide solutions useful in their own right for assessment of walking machine performance. Gait cycle simulations (i.e., combination of a leg recovery and a body move) are useful for evaluating mechanism performance issues such as stability, power consumption, traversability and potential for foot slippage. Three cases of gait cycle simulations are considered for different soil types and gravitational constants. Leveling control algorithms for maintaining AMBLER body tilt are tested through model simulations prior to implementation on the actual mechanism. The generality of the legged locomotion model is demonstrated through application to the Overlapping Walker configuration for flat settlement and body move simulations.

7.1 Gait Cycle Simulations

Initial gaits intended for the AMBLER walking machine are composed of discrete body move, leg lift, leg recovery and foot placement motions. Gait cycle simulations are conducted by specifying joint trajectories that are consistent with the mechanism motions shown in Table 7.1. Corresponding mechanism stances at the beginning and end of the discrete motions are illustrated in Figure 7.1. Body and joint trajectories for each discrete motion follow a quintic profile beginning and ending at zero velocity and acceleration. All shoulder and elbow joints are considered to be powered while the vertical joint positions are fixed at $1.0914\,m$. Physical properties for the AMBLER mechanism are those described in Section 5.3.

Three soil types are considered for the gait cycle simulations: compact sand, loose sand and Martian soil. The compact sand was used for the verification analyses of Chapter 6. Force deflection relationships augmented with structural compliance are given in Figure 6.1. The loose sand (which was used for the foot-soil interaction experiments described in Appendix A) and Martian soil

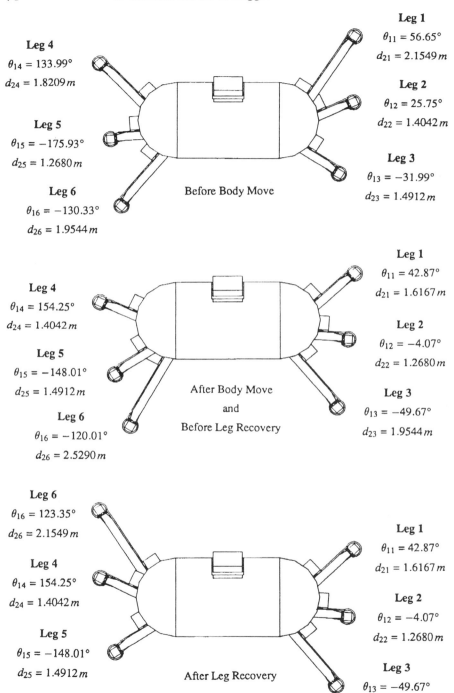

Figure 7.1: Mechanism Stances Used for Gait Cycle Simulations

Model Applications

Time Interval	Corresponding Mechanism Motions
$0 - 10\,sec$	70 cm body move
$10 - 14\,sec$	raise Leg 6 vertical member 30.48 cm
$14 - 18.96\,sec^{(1)}$	shorten Leg 6 elbow 151.38 cm
$18.96 - 24\,sec^{(1)}$	extend Leg 6 elbow 113.97 cm
$14 - 24\,sec^{(1)}$	rotate Leg 6 shoulder counterclockwise 243.36°
$24 - 28\,sec$	lower Leg 6 vertical member 29.21 cm

(1) Elbow and shoulder motions are coordinated so that the recovering leg is fully retracted as it passes through the body cavity.

Table 7.1: Gait Cycle Mechanism Motions

[53] are characterized by the force-deflection relationships (including structural compliance) shown in Figures 7.2 and 7.3.

A gravitational constant equal to Earth gravity, $9.81\,m/sec^2$, is used for the simulations on compact and loose sand while Mar's gravity, $3.78\,m/sec^2$, is used for the simulation on Martian soil. The local step error used in all analyses is 1×10^{-5} and the iteration error is 1×10^{-8}. The initial mechanism states are determined from settling analyses. Calculated results for gait cycle simulations on compact sand, loose sand, and Martian soil are given in Figures 7.4, 7.5, and 7.6, respectively. Motor winding losses, drivetrain inefficiencies, output power and amplifier losses are included in the calculated power results.

The general mechanism response is the same for all simulations. The body move and leg recovery motions affect the vertical foot forces due to a changing center of gravity. Foot placement and liftoff create load redistribution, impulsive loadings and high-frequency oscillations throughout the system.

The maximum input power is 19% greater during the body move on loose sand relative to compact sand because the loose sand experiences greater deformations. The input power required for leg recovery is nearly identical for loose and compact sands since these motions are relatively independent of soil type. Input power for the gait cycle simulation, utilizing Martian soil and gravity, is less than the other cases because the mechanism is effectively lighter. Foot liftoff and placement cause pulsations in the input power results for all cases.

7.2 Body Leveling Simulations

The AMBLER configuration and mechanism stance shown in Figure 7.7 are modeled for the body leveling simulations. Legs 1 and 2 are located on a rigid surface (i.e., only structural compliance is modeled at the foot-soil interface) while the remaining legs are situated on a loose sand (see Figure 7.2). The initial, tilted mechanism state is determined from a settling simulation and further tilting of

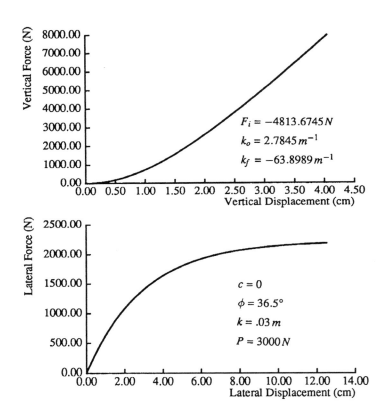

Figure 7.2: Force-Deflection Relationships for a Loose Sand

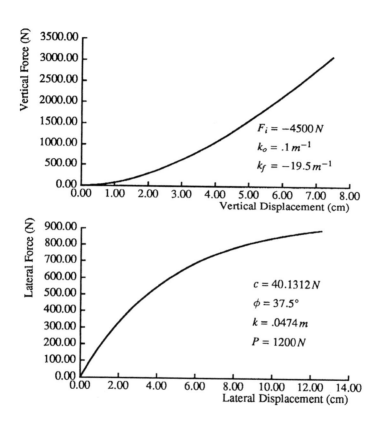

Figure 7.3: Force-Deflection Relationships for Martian Soil

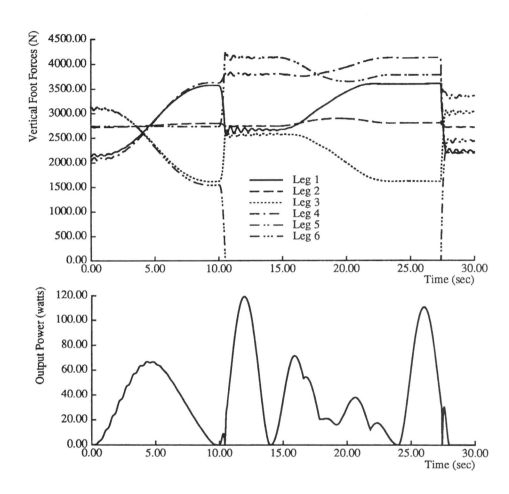

Figure 7.4: Results for Gait Cycle Simulation on a Compact Sand

Model Applications

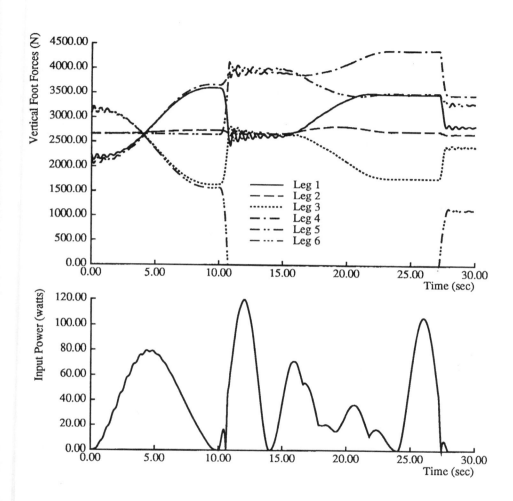

Figure 7.5: Results for Gait Cycle Simulation on a Loose Sand

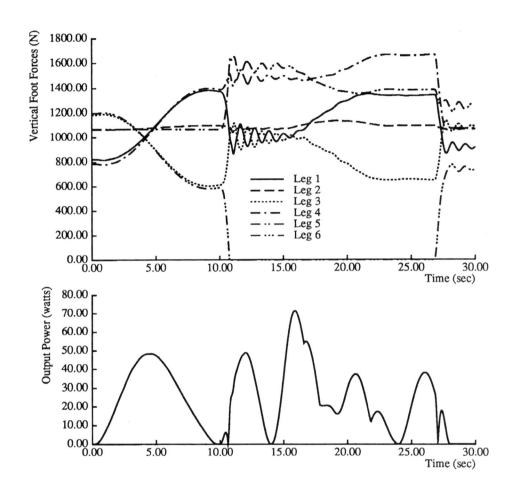

Figure 7.6: Results for Gait Cycle Simulation on Martian Soil

Model Applications

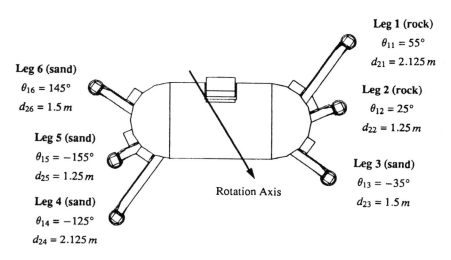

Figure 7.7: Mechanism Stance Used for Body Leveling Simulations

the mechanism by adjusting vertical leg lengths until a 3° body tilt (equivalent angle defined by roll-pitch-yaw rotations) is obtained. The axis of rotation for the tilted mechanism state is shown in Figure 7.7.

The desired body trajectory is a quintic profile beginning and ending at zero velocity and acceleration from the initial, tilted state to a level condition. This trajectory implicitly determines a corresponding set of vertical leg motions because all shoulder and elbow joints are fixed during the leveling operation. The leveling algorithm uses position and derivative control (i.e., feedback of joint position and velocity information) independently applied to the vertical leg axes for the determination of control forces without consideration of actuator saturation. Constant position and velocity gains are used for actively controlling all six legs.

The control forces are updated every 5 $msec$ to simulate a 200 Hz sampling frequency in the actual system. The complete leveling operation is considered to span 2 $secs$ and a constant stepsize of .001 sec is used for the simulations. This constant stepsize results in a local error less than the specified limit of 1×10^{-5} and the iteration error is 1×10^{-8}. Calculated results for a stable set of gains ($k_p = 500\ N/m$ and $k_d = 300\ N\ sec/m$) and a set of unstable gains ($k_p = 5000\ N/m$ and $k_d = 300\ N\ sec/m$) are given in Figures 7.8 and 7.9, respectively.

Results for the stable set of gains closely follow the desired leveling trajectory with relatively small variations in vertical foot forces. The mechanism center of gravity does not appreciably deviate for a 3° body tilt and vertical foot forces are expected to remain nearly constant during a well coordinated leveling operation. Increasing the position gain by an order of magnitude leads to control system instabilities which cause the mechanism to hop into the air and

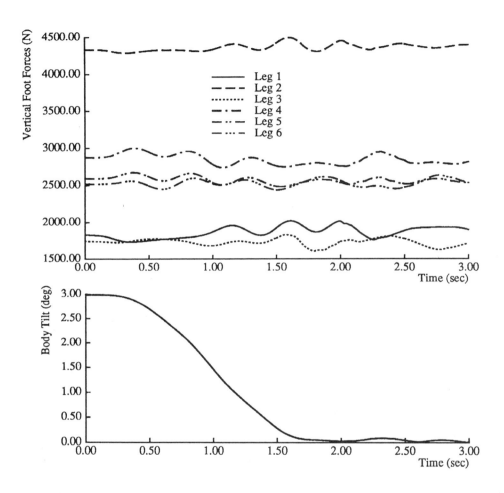

Figure 7.8: Body Leveling Results for a Set of Stable Gains

Model Applications

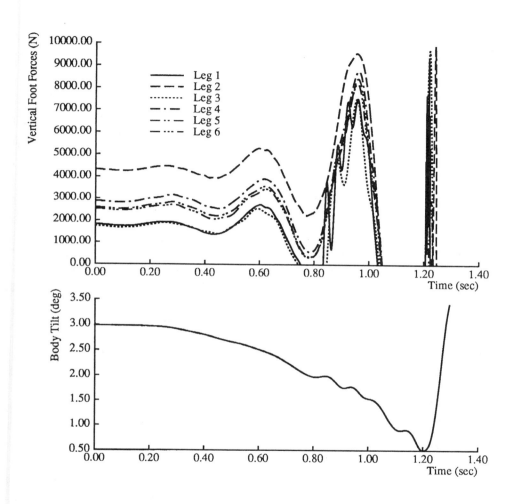

Figure 7.9: Body Leveling Results for a Set of Unstable Gains

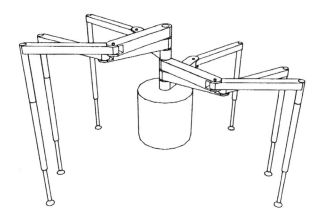

Figure 7.10: Overlapping Walker Configuration

recontact the ground. This example demonstrates the utility of the legged locomotion model and solution procedures for simulating accidental foot liftoff and freefall conditions. Additional examples of postural control simulations using the legged locomotion model on natural terrain are given in [56] and [55].

7.3 Example of Model Generality

The Overlapping Walker [46] shown in Figure 7.10 has six legs attached to a common vertical axis. Each leg can rotate fully around the vertical axis with a complete overlap of their workspaces. Two lateral members connected by revolute joints and a vertical telescoping member comprise a leg of this orthogonal legged walking machine. The Overlapping Walker is similar to the AMBLER configuration except for a rotational elbow joint and a single body post for the Overlapping design.

The same approximation is made in the Overlapping Walker geometric representation by considering all legs to attach at the same body post elevation, thus neglecting the vertical offset between legs. The assignment of coordinate frames is shown in Figure 7.11 with the corresponding Denavit-Hartenberg parameters given in Table 7.2. There are a total of 24 generalized coordinates. Global foot positions (used for calculating foot-soil interaction forces), generalized body and joint forces, and specialization of the dynamic equations are similar to the expressions developed for the AMBLER configuration in Sections 5.2 and 5.4. However, the U_{888} and U_{988} partial derivative matrices are non-zero for the Overlapping Walker configuration.

The pseudo-inertia matrices and center of mass vectors used for the Overlapping Walker configuration are given in Table 7.3 while the joint damping and backdriving parameters are specified in Table 7.4. All remaining model and so-

Model Applications

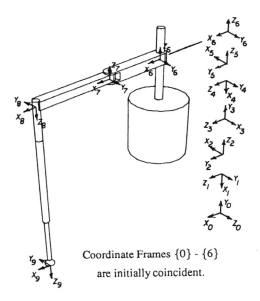

Coordinate Frames {0} - {6} are initially coincident.

Figure 7.11: Overlapping Walker Coordinate Frames

Link	θ	d	a	α
1	180°	δ_z	0	90°
2	−90°	δ_y	0	90°
3	−90°	δ_x	0	90°
4	$\theta_z - 90°$	0	0	90°
5	$\theta_y - 90°$	0	0	90°
6	$\theta_x + 90°$	0	0	0
7	θ_{1_i}	0	l_1	0
8	θ_{2_i}	0	l_2	180°
9	0	d_{3_i}	0	0

Table 7.2: Overlapping Walker Denavit-Hartenberg Parameters

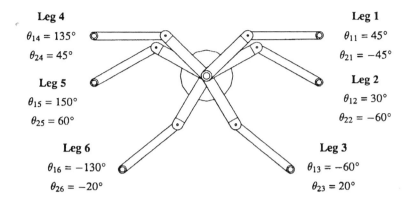

Figure 7.12: Mechanism Posture Used for Overlapping Walker Simulations

lution parameters are the same as those which were used for the verification studies considering the AMBLER configuration (refer to Section 6.1). The force-deflection relationships for a compact sand augmented with AMBLER structural compliances (see Figure 6.1) are used for the Overlapping Walker simulations which assumes that the two configurations have similar structural flexibilities.

A flat settlement example is analyzed for the symmetric mechanism posture shown in Figure 7.12 where all joint positions are fixed and vertical leg lengths are $2.9083\,m$. A $.001\,sec$ initial stepsize is used for this simulation which satisfies the local step error for the first timestep. Calculated results for the flat settlement example are shown in Figure 7.13.

A body move was simulated using the settled mechanism state as initial conditions. The desired motion is a $.4\,m$ translation of the body along the axis of symmetry for the mechanism stance. The body trajectory follows a quintic profile beginning and ending at zero velocity and acceleration which determines the shoulder and elbow joint trajectories that are applied as boundary conditions to the model. All lateral joints are powered and vertical joint positions are kept fixed at $2.9083\,m$. Calculated results for the body move example are given in Figure 7.14

The Overlapping Walker results are similar to the verification results in Chapter 6 for the AMBLER configuration. One notable difference is a larger amount of rocking of the Overlapping Walker mechanism during settlement and propulsion because of the higher center of gravity. These examples demonstrate the relative ease of analyzing alternate configurations with the legged locomotion model developed in this monograph. Consideration of alternate configurations simply requires redefinition of the Denavit-Hartenberg parameters and physical properties for the mechanism.

$$\mathbf{J}_6 = \begin{bmatrix} 50. & 0 & 0 & 0 \\ 0 & 50. & 0 & 0 \\ 0 & 0 & 1275.6517 & 998.3893 \\ 0 & 0 & 998.3893 & 800. \end{bmatrix} \quad \bar{\mathbf{r}}_6 = \begin{bmatrix} 0 \\ 0 \\ 1.2480 \\ 1 \end{bmatrix}$$

$$\mathbf{J}_7 = \begin{bmatrix} 23.4833 & -.1507 & -.1861 & -28.3048 \\ -.1507 & .2205 & .0038 & .1404 \\ -.1861 & .0038 & .1205 & .0998 \\ -28.3048 & .1404 & .0998 & 43.1649 \end{bmatrix} \quad \bar{\mathbf{r}}_7 = \begin{bmatrix} -.6557 \\ .0033 \\ .0023 \\ 1 \end{bmatrix}$$

$$\mathbf{J}_8 = \begin{bmatrix} 9.9535 & 0 & 0 & -12.9811 \\ 0 & -2.0342 & 0 & 0 \\ 0 & 0 & 136.7550 & 71.5339 \\ -12.9811 & 0 & 71.5339 & 72.0933 \end{bmatrix} \quad \bar{\mathbf{r}}_8 = \begin{bmatrix} -.1801 \\ 0 \\ .9922 \\ 1 \end{bmatrix}$$

$$\mathbf{J}_9 = \begin{bmatrix} .0046 & 0 & 0 & 0 \\ 0 & .0046 & 0 & 0 \\ 0 & 0 & 91.6060 & -45.4711 \\ 0 & 0 & -45.4711 & 31.9652 \end{bmatrix} \quad \bar{\mathbf{r}}_9 = \begin{bmatrix} 0 \\ 0 \\ -1.4225 \\ 1 \end{bmatrix}$$

$$\mathbf{J}_i = \begin{bmatrix} kg\,m^2 & kg\,m^2 & kg\,m^2 & kg\,m \\ kg\,m^2 & kg\,m^2 & kg\,m^2 & kg\,m \\ kg\,m^2 & kg\,m^2 & kg\,m^2 & kg\,m \\ kg\,m & kg\,m & kg\,m & m \end{bmatrix} \quad \bar{\mathbf{r}}_i = \begin{bmatrix} m \\ m \\ m \\ 1 \end{bmatrix}$$

All pseudo inertia terms are calculated to three significant figures while all cg locations are calculated to four figures.

Table 7.3: Overlapping Walker Physical Properties

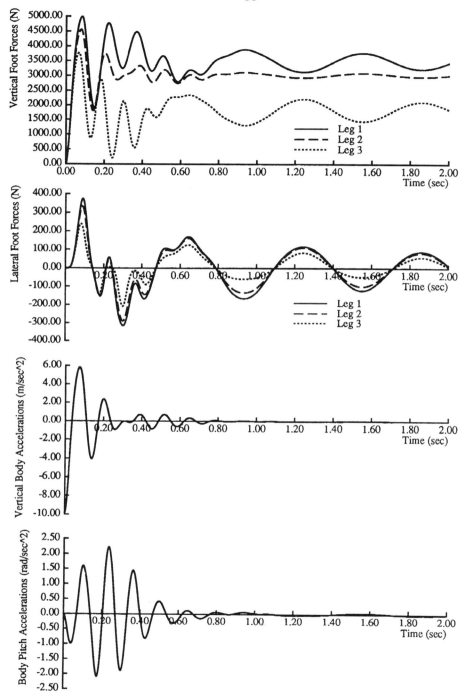

Figure 7.13: Overlapping Walker Flat Settlement Results

Model Applications

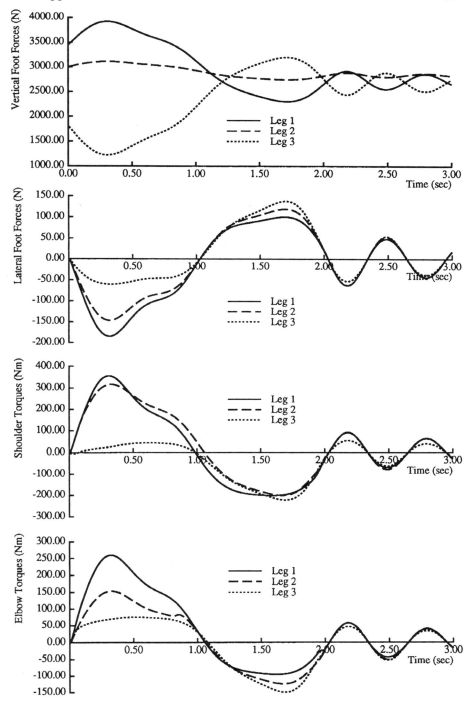

Figure 7.14: Overlapping Walker Body Move Results

Joint	Coulomb Friction Value[1] (N or Nm)	Asymptotic Backdriving Force (N or Nm)
shoulder	15 *or* 28	52
elbow	8 *or* 16	20
vertical	6 *or* 12	15

(1) The higher friction values apply if a joint moment loading is generated by foot-soil interactions.

Table 7.4: Overlapping Walker Joint Damping and Backdriving Parameters

Chapter 8

Summary

8.1 Summary

This research develops a dynamic mechanism model that characterizes indeterminate interactions of a closed-chain robot with its environment. The general approach is to identify and model the contact compliance, then define a set of generalized coordinates that are advantageous for this class of robot. Legged locomotion on natural terrain is modeled to illustrate the methodology, although the approach is applicable to any dynamic, closed-chain mechanism with sufficient contact compliance.

The set of generalized coordinates comprised of six body motions plus all appendage joint motions is shown to exhibit distinct advantages for representing legged locomotion on natural terrain. This coordinate set implicitly accommodates the mechanism closed-chains and the resulting model does not require reformulation each time a ground contact is established or broken. Therefore, the legged locomotion model developed in this work is ideally suited for simulating tipover, freefall and foot liftoff/placement. Lagrangian dynamics combined with homogeneous transformation matrices are used to derive equations of motion that can be applied to any walking machine configuration (e.g., bipeds, quadrupeds, beam-walkers and hopping machines) by specifying appropriate geometric and physical parameters, without reformulating the model in any way.

Compliant foot-soil interactions are modeled with empirical force-deflection equations for different terrain materials, slopes and loading conditions. Force-deflection equations were adopted from terramechanics methods. The bilinear equation used for modeling vertical foot-soil interactions is partially qualified with experimental results for walking machine applications. Similarly, joint damping and backdriving are modeled with empirical equations that are supported by experimental results.

Solution procedures have been developed that are especially suited for solving the differential/algebraic equations (DAE's) that characterize a model of legged locomotion on natural terrain. All problem variations arising from different combinations of specified joint force histories and joint trajectories are solved in the same manner by applying an Euler predictor-corrector algorithm to the original set of equations. This approach is efficient, robust and maximizes the versatility of the solution procedures. Automatically variable stepsizes are used to control the global error and to increase solution efficiency. Impulsive loads from foot liftoff/placement are accommodated by automatic stepsize reduction.

The algebraic equations that result from discretizing the equations of motion are solved using a combination of full Newton, altered Newton and chord iterative solution methods. The full Newton method is used for the first iteration while the altered Newton approach is used for subsequent iterations. Minimal variations were observed in the computationally intensive dynamic contributions to the Jacobian. The novel altered Newton method exploits this behavior by updating only the relatively quick, yet more significant, joint and foot force contributions. The chord method provides additional computational savings because the Jacobian is held constant between iterations and does not require decomposition. Solution convergence is monitored and the iterative solution switches from the altered Newton to the chord method once the solution has stabilized. A difference approximation is used to calculate the Jacobian for the system equations because unloading characteristics and intermittent contact of the foot-soil modeling preclude symbolic evaluation.

Modeling and solution procedures are demonstrated through application of the legged locomotion model to the AMBLER prototype walking machine. The equations of motion are specialized to take advantage of the orthogonal legged configuration for a significant reduction in computational effort. A massless leg model is formulated where the leg inertias are reallocated to the body resulting in two orders of magnitude reduction in run times compared to the comprehensive model. The simplified model is sufficiently accurate for use in preliminary investigations of a mechanism configuration.

Simulations that isolate responses of the legged locomotion model are used to verify the modeling and solution procedures. Flat settlement, sloped settlement, body move and leg recovery motions are simulated thus accounting for all primitive mechanism motions that constitute walking on natural terrain. The calculated results for these cases are evaluated for physical reasonableness and for satisfaction of equilibrium and symmetry conditions.

The utility of the legged locomotion model on natural terrain is demonstrated by gait cycle and leveling control simulations. The gait cycle simulations estimate input power consumption for different soil types and gravitational constants. Simulations using the legged locomotion model for the development of leveling control algorithms for the AMBLER mechanism are examined. The generality of the legged locomotion model is illustrated through application to the Overlapping Walker configuration for flat settlement and body move simulations, which simply requires redefinition of Denavit-Hartenberg parameters and physical properties.

8.2 Future Research

As with any pioneering effort, an agenda of relevant extensions are now evident. These extensions span the areas of model formulation, solution procedures and terrain interactions. Kane's dynamical equations could be used to derive more efficient equations of motion for a specific configuration, but at the expense of model generality which is the hallmark of the current work. Joint and member flexibilities can be included in the current dynamic formulation to completely describe the system compliance. The set of generalized coordinates would include the six body motions, all appendage joint motions, plus all deflection variables that characterize the member and joint compliance. Incorporating these additional compliances significantly increases the number of dofs and, therefore, the computational effort; solution procedures would remain unchanged.

The equations of motion are not a set of stiff differential equations at all times (e.g., freefall, near steady-state), so computational performance could be improved by switching between implicit and explicit solution procedures as dictated by evaluation of the Jacobian. Similarly, using an altered Newton or chord iterative solution across multiple timesteps would minimize expensive evaluation of the dynamic equations. The convergence ratio would require monitoring to determine degradation of solution performance.

The largest uncertainty in modeling legged locomotion on natural terrain is the fidelity of the foot-soil interaction models. Open issues include the dependence of foot-soil interactions on soil slope, sole design, rotational loading, and numerous other factors. In this work, an exponential transition to Coulomb's equation is used to model lateral foot-soil interactions but this representation assumes a constant vertical load and rectilinear motions. Neither of these conditions actually exist for legged locomotion and approximations were made to facilitate model formulation. Extensive research is required to understand and quantify foot-soil interactions and their dependence on configuration and terrain parameters.

8.3 Perspectives

The proposed goal of this work was to develop a functional model of legged locomotion on natural terrain. Specific objectives were to incorporate realistic foot-soil interactions and to produce solutions suitable for real-time applications. A general model was not envisioned and a reformulation would be required for each walker configuration.

The original approach was to formulate the equations of motion using Lagrangian dynamics and apply constraint equations for representation of closed-chains. This led to a large set of system equations (approximately 100) for a hexapod walker; pre-computation of dynamic coefficients was intended to make the approach suitable for real-time applications, but this was naive given current computing capabilities. Because constraint equations were to be used, the resulting equations of motion would be coupled DAE's which require relatively sophisticated solution techniques.

While deriving the enormous number of system equations, it became apparent that real-time solution of these equations would be prohibitive and their subsequent verification would be difficult. Also, it became obvious that implicit solution techniques (i.e., iterative) were required for DAE systems, further complicating the modeling and solution procedures.

During the formulation of the equations of motion for the Overlapping Walker configuration, the immense benefit of the set of generalized coordinates used in this work was realized. The advantages of this revised approach include a reduction to 24 dofs for any hexapod walker, foot placement/liftoff is directly accommodated, and the resulting model generalizes to all configurations. Also, the equations of motion become an uncoupled set of DAE's, permitting simpler solution procedures.

Lagrangian dynamics were still used to derive the equations of motion resulting in a general, but computationally intensive, formulation. However, real-time model applications were still not possible. The decision was made to emphasize the generality, robustness and utility of the legged locomotion model and corresponding solution methods. The legged locomotion model, obtained using Lagrangian dynamics, is applicable to all configurations (e.g., bipeds, quadrupeds, beam-walkers and hopping machines) through suitable definition of geometric parameters. Once a configuration is fixed, a more computationally efficient model could be formulated for a specific configuration using Kane's dynamics. The set of generalized coordinates and solution procedures would remain unchanged.

Novel solution procedures were devised for this class of problem, permitting the efficient calculation of stable, accurate solutions and adaptability to all problem variations arising from different combinations of imposed boundary conditions. An extremely robust, numerical scheme was implemented because the model must apply to extreme events, such as accidental foot contacts. If the computational performance of the existing solution procedures becomes limiting, enhanced algorithms such as solution switching, application of altered Newton and chord methods across multiple timesteps, and parallel processing can be developed without model reformulation.

The legged locomotion model was applied to a prototype walking machine, and the simulations generated significant insights into walking machine performance on natural terrain. Current uses of the model include evaluation of stability margins against tipover and development of propulsion control algorithms for the AMBLER configuration. Near term model applications entail detailed performance assessments (e.g., power consumption, maximum grade of traversal, sensitivity to soil type) that will provide critical insights for incorporation into the AMBLER motion planner.

This work develops a functional model of legged locomotion that incorporates, for the first time, non-conservative foot-soil interactions into a non-linear dynamic formulation. The resulting model is general for all walker configura-

Summary

tions and corresponding solution procedures are extremely robust and accommodate all problem variations. The studies of this work, which illustrate the methodology, are original and essential contributions to the design, evaluation and control of these complex robot systems.

Appendix A

Vertical Foot-Soil Experiments

Vertical foot placement experiments were conducted with the Overlapping Walker Single Leg Testbed [52] shown in Figures 3.8 and 3.10 to qualify the bilinear equation (Equation 2.5) for modeling vertical foot-soil interactions. The vertical leg member was braced against a building girder to virtually eliminate link deflections from the experiments. Vertical foot positions were obtained from joint encoder readings and six foot force components (i.e., three forces and three moments) were measured.

A common highway sand with shear properties of $c = 0$ and $\phi = 36.5°$ was used for the tests. Soil slopes of 0°, 10°, 20°, and 30° were investigated and three tests were conducted for each slope condition. The sand was completely re-mixed between tests to reverse the effects of compaction. A $0 - 4500\,N$ load range was investigated which corresponds with the expected loading of the AMBLER prototype walking machine. Data was taken approximately at each $225\,N$ increment of loading. The foot force sensor was oriented relative to the sloped surface during loading such that single components of lateral force and bending moment were developed as shown in Figure A.1.

The results of the three tests for each slope condition were normalized so that vertical displacements would be equal for a $225\,N$ vertical load (or the nearest data point). The results were normalized in this manner because variations in surface conditions affected the onset of loading for each test, thus shifting the load-deflection curves. These variations between tests were effectively eliminated when the vertical load reached $225\,N$.

Experimental results, along with fitted bilinear curves for the different slope conditions, are given in Figures A.2 - A.5; parameters for the bilinear curves were calculated using a least-squares fit. In general, there is good correlation between the sets of results for each slope condition. However, there is an offset in a set of results for the 10° slope case (Figure A.3) attributed to inconsistent soil preparation between experiments.

98 A General Model of Legged Locomotion on Natural Terrain

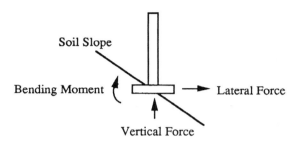

Figure A.1: Measured Foot Force Components

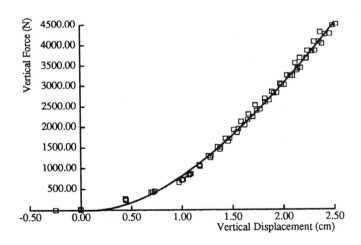

Figure A.2: Experimental Results for 0° Ground Slope

Vertical Foot-Soil Experiments

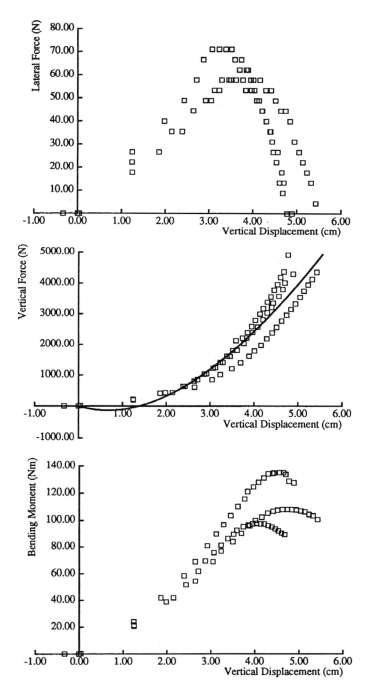

Figure A.3: Experimental Results for 10° Ground Slope

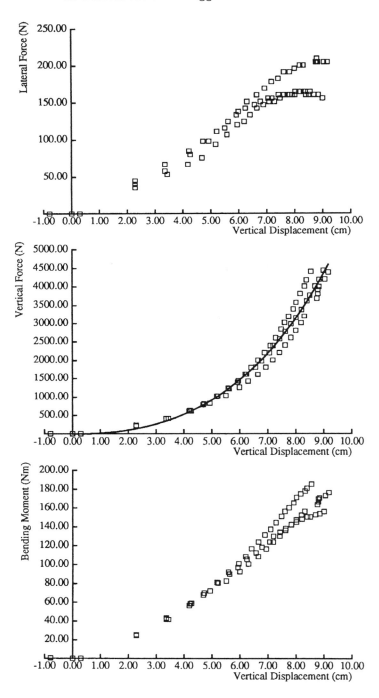

Figure A.4: Experimental Results for 20° Ground Slope

Vertical Foot-Soil Experiments 101

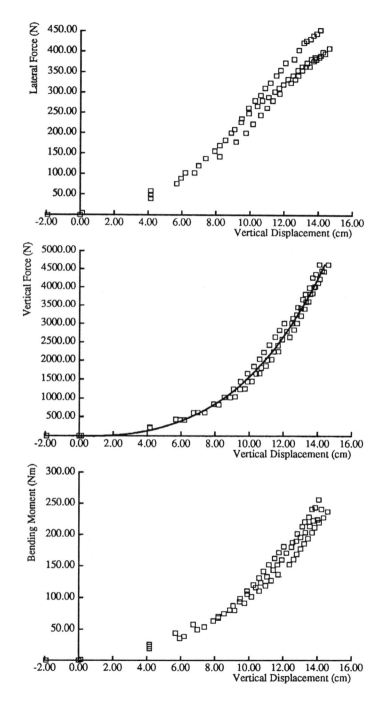

Figure A.5: Experimental Results for 30° Ground Slope

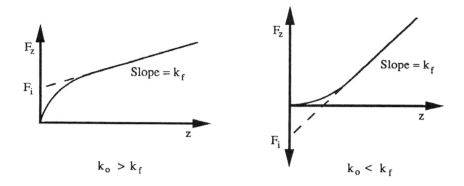

Figure A.6: Graphical Interpretation of Bilinear Equation Parameters

As the ground slope increases, larger lateral forces and bending moments are developed, and the effective vertical compliance of the soil increases. It appears that lateral forces and bending moments approach maximum values for higher sinkages on sloped surfaces. The lateral forces and bending moments for the 10° slope case (Figure A.3) decrease after the maximum values have been attained. These loads are relieved because the sloped surface flattens locally during foot placement, thus reducing the asymmetric stress distribution on the foot. Extensive research is required to understand foot placement on sloped surfaces and the complex loading conditions that are involved.

The bilinear equation affords an excellent representation of vertical foot-soil interactions for flat and sloped surfaces. Caution must be exercised when fitting bilinear equation parameters because the resulting function may exhibit a tensile response between the foot and soil at lower sinkage values as shown by the 10° slope results (Figure A.3). This anomaly is only possible for soils having an initial slope less than the final slope (i.e., $k_o < k_f$).

If this erroneous representation occurs, the bilinear equation parameters should be adjusted by the analyst's judgment as recommended in [24] for all instances of curve fitting to soil data. These parameters are particularly amenable for fitting by "eye" because of their graphical interpretation as shown in Figure A.6 for two soil types. Intersection of the plastic loading line with the load axis determines the axis intercept F_i, while the slope of the plastic loading line is equal to $F_i k_f$. The final parameter k_o is adjusted so that the transition region from initial to plastic loading is adequately represented.

Appendix B

Perturbation Parameters

The solution Jacobian is formulated using a finite difference approximation (Equation 4.5) which requires two sets of parameters specifying the perturbation imposed for each dof.

$$\frac{\partial \mathbf{J}}{\partial y_j} = \frac{\mathbf{f}^* - \mathbf{f}}{\delta} \tag{B.1}$$

where $\frac{\partial \mathbf{J}}{\partial y_j}$ - j^{th} column of the Jacobian,
$\mathbf{f}$ - residual vector for trial solution,
$\mathbf{f}^*$ - residual vector for perturbed solution,
$\mathbf{y}^*$ - perturbed solution = $(y_1, y_2 \ldots y_j + \delta \ldots y_n)$, and
δ - perturbation of coordinate j.

The first set of parameters prescribes relative perturbations for dofs with non-zero positions (i.e., greater than .01) while absolute perturbations for dofs with near-zero positions are specified by the second set of parameters. (A relative perturbation applied to a zero position results in no change at all.) The choice of perturbation parameters affects the stability of the Jacobian and consequently influences the convergence of the iterative solution.

Convergence sensitivity to the perturbation parameters was determined through trial simulations of the AMBLER configuration during settlement on flat compact sand (Case 1 verification model of Section 6.2). Parameters were adjusted and the corresponding solution convergence was observed. A constant stepsize of .001 sec was used for the simulations to prevent the solution algorithm from reducing the stepsize to compensate for poor convergence.

Convergence ratios for a few of the investigated cases are shown in Table B.1; results are presented for the first timesteps of these runs. The convergence ratio is defined as the ratio of successive altered Newton iteration errors immediately before the solution switches to the chord method (i.e., the Jacobian has stabilized). For Cases A and B, the solution does not converge (i.e., satisfy the iteration error within 10 iterations) after 18 timesteps. Optimum convergence was obtained with the parameters of Case E and these parameters are used

	Perturbation Parameters						
	Relative			Absolute			
Case	BT	BR	AM	BT	BR	AM	CR
A	.01	.005	.01	.0001	.0001	.01	.2386
B	.01	.005	.01	.001	.0001	.01	.2136
C	.01	.005	.01	.05	.01	.01	.1626
D	.01	.005	.01	.5	.1	.1	.1623
E	.05	.025	.05	.5	.1	.1	.0819

BT - body translations (dofs 1-3)[1]
BR - body rotations (dofs 4-6)[1]
AM - appendage motions (dofs 7-9)[1]
CR - convergence ratio[2]

(1) The dof numbering refers to AMBLER coordinates illustrated in Figure 5.4.
(2) Defined as the ratio of successive altered Newton iteration errors immediately before the solution switches to the chord method.

Table B.1: Convergence Ratios for Different Sets of Perturbation Parameters

with excellent success for all of the numerical computations presented in this work.

Appendix C

Inertial Properties for the AMBLER Massless Leg Model

A simplified model of the AMBLER mechanism is formulated by reallocating the leg inertias to the body at the beginning of each timestep. The parallel axis theorem cannot be used in this case for transforming leg properties because axes of the leg members are normally not parallel to the body axes. Leg properties are transformed by explicitly defining positions of a leg member relative to the body axes and substituting these expressions into the body inertial equations for subsequent integration. The modified body inertial properties are obtained by combining all leg member contributions with the original body properties.

Integration of the body inertial equations (after substitution of the relative positions for a leg member) is simplified by recognizing that each term can be expressed as functions of pseudo-inertia matrix terms for the leg member being considered. The pseudo-inertia matrix for a rigid body is given below.

$$\mathbf{J} = \begin{bmatrix} \frac{I_{yy}+I_{zz}-I_{xx}}{2} & I_{xy} & I_{xz} & mx \\ I_{xy} & \frac{I_{xx}+I_{zz}-I_{yy}}{2} & I_{yz} & my \\ I_{xz} & I_{yz} & \frac{I_{xx}+I_{yy}-I_{zz}}{2} & mz \\ mx & my & mz & m \end{bmatrix}$$

$$= \begin{bmatrix} \int x^2 dm & \int xy\, dm & \int xz\, dm & \int x\, dm \\ \int xy\, dm & \int y^2 dm & \int yz\, dm & \int y\, dm \\ \int xz\, dm & \int yz\, dm & \int z^2 dm & \int z\, dm \\ \int x\, dm & \int y\, dm & \int z\, dm & \int dm \end{bmatrix}$$

Therefore, reallocation of leg inertial properties is characterized by identifying simple functions of the original leg properties and extensive symbolic evaluations are not required. The $^6\mathbf{A}_7$, $^7\mathbf{A}_8$ and $^8\mathbf{A}_9$ coordinate frames, shown in Figure 5.4 with the corresponding Denavit-Hartenberg parameters given in Table 5.1, are used to define the leg member positions relative to the body coordinate axes. In the following derivations, all primed expressions represent contributions of a leg member to the body parameters.

C.1 Reallocation of Inner Link Properties

Positions of an inner link relative to the body axes are:

$$x_6 = -x_7 \sin\theta_{1_i} + z_7 \cos\theta_{1_i} + dl_i$$
$$y_6 = x_7 \cos\theta_{1_i} + z_7 \sin\theta_{1_i}$$
$$z_6 = y_7$$

Substitution of these expressions into the body inertial equations and subsequent evaluation results in the following contributions of an inner link to the body properties.

$$(I_{xx})'_{inner_i} = \mathbf{J}_7(1,1)\cos^2\theta_{1_i} + \mathbf{J}_7(2,2) + \mathbf{J}_7(3,3)\sin^2\theta_{1_i}$$
$$+ 2\mathbf{J}_7(1,3)\cos\theta_{1_i}\sin\theta_{1_i}$$

$$(I_{yy})'_{inner_i} = \mathbf{J}_7(1,1)\sin^2\theta_{1_i} + \mathbf{J}_7(2,2) + \mathbf{J}_7(3,3)\cos^2\theta_{1_i} + 2\mathbf{J}_7(3,4)dl_i\cos\theta_{1_i}$$
$$- 2\mathbf{J}_7(1,3)\cos\theta_{1_i}\sin\theta_{1_i} - 2\mathbf{J}_7(1,4)dl_i\sin\theta_{1_i} + \mathbf{J}_7(4,4)dl_i^2$$

$$(I_{zz})'_{inner_i} = \mathbf{J}_7(1,1) + \mathbf{J}_7(3,3) + \mathbf{J}_7(4,4)dl_i^2 - 2\mathbf{J}_7(1,4)dl_i\sin\theta_{1_i}$$
$$+ 2\mathbf{J}_7(3,4)dl_i\cos\theta_{1_i}$$

$$(I_{xy})'_{inner_i} = -\mathbf{J}_7(1,1)\cos\theta_{1_i}\sin\theta_{1_i} + \mathbf{J}_7(3,3)\cos\theta_{1_i}\sin\theta_{1_i}$$
$$+ \mathbf{J}_7(1,3)\cos 2\theta_{1_i} + \mathbf{J}_7(1,4)dl_i\cos\theta_{1_i} + \mathbf{J}_7(3,4)dl_i\sin\theta_{1_i}$$

$$(I_{xz})'_{inner_i} = -\mathbf{J}_7(1,2)\sin\theta_{1_i} + \mathbf{J}_7(2,3)\cos\theta_{1_i} + \mathbf{J}_7(2,4)dl_i$$

$$(I_{yz})'_{inner_i} = \mathbf{J}_7(1,2)\cos\theta_{1_i} + \mathbf{J}_7(2,3)\sin\theta_{1_i}$$

$$(mx)'_{inner_i} = -\mathbf{J}_7(1,4)\sin\theta_{1_i} + \mathbf{J}_7(3,4)\cos\theta_{1_i} + \mathbf{J}_7(4,4)dl_i$$

$$(my)'_{inner_i} = \mathbf{J}_7(1,4)\cos\theta_{1_i} + \mathbf{J}_7(3,4)\sin\theta_{1_i}$$

$$(mz)'_{inner_i} = \mathbf{J}_7(2,4)$$

Inertial Properties for the AMBLER Massless Leg Model

C.2 Reallocation of Outer Link Properties

Positions of an outer link relative to the body axes are:

$$x_6 = -x_8 \sin\theta_{1_i} + y_8 \cos\theta_{1_i} + d_{2_i} \cos\theta_{1_i} + dl_i$$
$$y_6 = x_8 \cos\theta_{1_i} + y_8 \sin\theta_{1_i} + d_{2_i} \sin\theta_{1_i}$$
$$z_6 = -z_8$$

Substitution of these expressions into the body inertial equations and subsequent evaluation results in the following contributions of an outer link to the body properties.

$$(I_{xx})'_{outer_i} = \mathbf{J}_8(1,1)\cos^2\theta_{1_i} + \mathbf{J}_8(2,2)\sin^2\theta_{1_i} + \mathbf{J}_8(3,3) + \mathbf{J}_8(4,4)d_{2_i}^2 \sin^2\theta_{1_i}$$
$$+ 2\mathbf{J}_8(1,2)\cos\theta_{1_i}\sin\theta_{1_i} + 2\mathbf{J}_8(1,4)d_{2_i}\cos\theta_{1_i}\sin\theta_{1_i}$$
$$+ 2\mathbf{J}_8(2,4)d_{2_i}\sin^2\theta_{1_i}$$

$$(I_{yy})'_{outer_i} = \mathbf{J}_8(4,4)(d_{2_i}\cos\theta_{1_i} + dl_i)^2 - 2\mathbf{J}_8(1,2)\cos\theta_{1_i}\sin\theta_{1_i} + \mathbf{J}_8(3,3)$$
$$- 2\mathbf{J}_8(1,4)(d_{2_i}\cos\theta_{1_i} + dl_i)\sin\theta_{1_i} + \mathbf{J}_8(1,1)\sin^2\theta_{1_i}$$
$$+ 2\mathbf{J}_8(2,4)(d_{2_i}\cos\theta_{1_i} + dl_i)\cos\theta_{1_i} + \mathbf{J}_8(2,2)\cos^2\theta_{1_i}$$

$$(I_{zz})'_{outer_i} = \mathbf{J}_8(1,1) + \mathbf{J}_8(2,2) + \mathbf{J}_8(4,4)(d_{2_i}^2 + dl_i^2 + 2d_{2_i}dl_i\cos\theta_{1_i})$$
$$- 2\mathbf{J}_8(1,4)dl_i\sin\theta_{1_i} + 2\mathbf{J}_8(2,4)(d_{2_i} + dl_i\cos\theta_{1_i})$$

$$(I_{xy})'_{outer_i} = \mathbf{J}_8(1,4)(d_{2_i}\cos 2\theta_{1_i} + dl_i\cos\theta_{1_i}) - \mathbf{J}_8(1,1)\cos\theta_{1_i}\sin\theta_{1_i}$$
$$+ \mathbf{J}_8(2,4)(d_{2_i}\sin 2\theta_{1_i} + dl_i\sin\theta_{1_i}) + \mathbf{J}_8(2,2)\cos\theta_{1_i}\sin\theta_{1_i}$$
$$+ \mathbf{J}_8(4,4)(d_{2_i}\cos\theta_{1_i} + dl_i)d_{2_i}\sin\theta_{1_i} + \mathbf{J}_8(1,2)\cos 2\theta_{1_i}$$

$$(I_{xz})'_{outer_i} = \mathbf{J}_8(1,3)\sin\theta_{1_i} - \mathbf{J}_8(2,3)\cos\theta_{1_i} - \mathbf{J}_8(3,4)(d_{2_i}\cos\theta_{1_i} + dl_i)$$

$$(I_{yz})'_{outer_i} = -\mathbf{J}_8(1,3)\cos\theta_{1_i} - \mathbf{J}_8(2,3)\sin\theta_{1_i} - \mathbf{J}_8(3,4)d_{2_i}\sin\theta_{1_i}$$

$$(mx)'_{outer_i} = -\mathbf{J}_8(1,4)\sin\theta_{1_i} + \mathbf{J}_8(2,4)\cos\theta_{1_i} + \mathbf{J}_8(4,4)(d_{2_i}\cos\theta_{1_i} + dl_i)$$

$$(my)'_{outer_i} = \mathbf{J}_8(1,4)\cos\theta_{1_i} + \mathbf{J}_8(2,4)\sin\theta_{1_i} + \mathbf{J}_8(4,4)d_{2_i}\sin\theta_{1_i}$$

$$(mz)'_{outer_i} = -\mathbf{J}_8(3,4)$$

C.3 Reallocation of Vertical Link Properties

Positions of a vertical link relative to the body axes are:

$$x_6 = x_9 \cos\theta_{1_i} + y_9 \sin\theta_{1_i} + d_{2_i} \cos\theta_{1_i} + dl_i$$
$$y_6 = x_9 \sin\theta_{1_i} - y_9 \cos\theta_{1_i} + d_{2_i} \sin\theta_{1_i}$$
$$z_6 = -z_9 - d_{3_i}$$

Substitution of these expressions into the body inertial equations and subsequent evaluation results in the following contributions of a vertical link to the body properties.

$$(I_{xx})'_{vert_i} = \mathbf{J}_9(1,1)\sin^2\theta_{1_i} + \mathbf{J}_9(2,2)\cos^2\theta_{1_i} + \mathbf{J}_9(3,3) + 2\mathbf{J}_9(3,4)d_{3_i}$$
$$+ \mathbf{J}_9(4,4)\left(d_{2_i}^2\sin^2\theta_{1_i} + d_{3_i}^2\right) - 2\mathbf{J}_9(1,2)\cos\theta_{1_i}\sin\theta_{1_i}$$
$$+ 2\mathbf{J}_9(1,4)d_{2_i}\sin^2\theta_{1_i} - 2\mathbf{J}_9(2,4)d_{2_i}\cos\theta_{1_i}\sin\theta_{1_i}$$

$$(I_{yy})'_{vert_i} = \mathbf{J}_9(1,1)\cos^2\theta_{1_i} + \mathbf{J}_9(2,2)\sin^2\theta_{1_i} + \mathbf{J}_9(3,3) + 2\mathbf{J}_9(3,4)d_{3_i}$$
$$+ \mathbf{J}_9(4,4)[(d_{2_i}\cos\theta_{1_i} + dl_i)^2 + d_{3_i}^2] + 2\mathbf{J}_9(1,2)\cos\theta_{1_i}\sin\theta_{1_i}$$
$$+ 2\mathbf{J}_9(1,4)(d_{2_i}\cos\theta_{1_i} + dl_i)\cos\theta_{1_i}$$
$$+ 2\mathbf{J}_9(2,4)(d_{2_i}\cos\theta_{1_i} + dl_i)\sin\theta_{1_i}$$

$$(I_{zz})'_{vert_i} = \mathbf{J}_9(1,1) + \mathbf{J}_9(2,2) + \mathbf{J}_9(4,4)\left(d_{2_i}^2 + dl_i^2 + 2d_{2_i}dl_i\cos\theta_{1_i}\right)$$
$$+ 2\mathbf{J}_9(1,4)(d_{2_i} + dl_i\cos\theta_{1_i}) + 2\mathbf{J}_9(2,4)dl_i\sin\theta_{1_i}$$

$$(I_{xy})'_{vert_i} = \mathbf{J}_9(1,1)\cos\theta_{1_i}\sin\theta_{1_i} + \mathbf{J}_9(4,4)d_{2_i}\sin\theta_{1_i}(d_{2_i}\cos\theta_{1_i} + dl_i)$$
$$- \mathbf{J}_9(1,2)\cos 2\theta_{1_i} + \mathbf{J}_9(1,4)(d_{2_i}\sin 2\theta_{1_i} + dl_i\sin\theta_{1_i})$$
$$- \mathbf{J}_9(2,4)(d_{2_i}\cos 2\theta_{1_i} + dl_i\cos\theta_{1_i}) - \mathbf{J}_9(2,2)\cos\theta_{1_i}\sin\theta_{1_i}$$

$$(I_{xz})'_{vert_i} = -\mathbf{J}_9(1,3)\cos\theta_{1_i} - \mathbf{J}_9(3,4)(d_{2_i}\cos\theta_{1_i} + dl_i) - \mathbf{J}_9(1,4)d_{3_i}\cos\theta_{1_i}$$
$$- \mathbf{J}_9(2,4)d_{3_i}\sin\theta_{1_i} - \mathbf{J}_9(4,4)d_{3_i}(d_{2_i}\cos\theta_{1_i} + dl_i)$$
$$- \mathbf{J}_9(2,3)\sin\theta_{1_i}$$

$$(I_{yz})'_{vert_i} = -\mathbf{J}_9(1,3)\sin\theta_{1_i} + \mathbf{J}_9(2,3)\cos\theta_{1_i} - \mathbf{J}_9(3,4)d_{2_i}\sin\theta_{1_i}$$
$$- \mathbf{J}_9(1,4)d_{3_i}\sin\theta_{1_i} + \mathbf{J}_9(2,4)d_{3_i}\cos\theta_{1_i} - \mathbf{J}_9(4,4)d_{2_i}d_{3_i}\sin\theta_{1_i}$$

$$(mx)'_{vert_i} = \mathbf{J}_9(1,4)\cos\theta_{1_i} + \mathbf{J}_9(2,4)\sin\theta_{1_i} + \mathbf{J}_9(4,4)(d_{2_i}\cos\theta_{1_i} + dl_i)$$

$$(my)'_{vert_i} = \mathbf{J}_9(1,4)\sin\theta_{1_i} - \mathbf{J}_9(2,4)\cos\theta_{1_i} + \mathbf{J}_9(4,4)d_{2_i}\sin\theta_{1_i}$$

$$(mz)'_{vert_i} = -\mathbf{J}_9(3,3) - \mathbf{J}_9(4,4)d_{3_i}$$

Bibliography

[1] A. A. Frank and R. B. McGhee, "Some Considerations Relating to the Design of Autopilots for Legged Vehicles," *Journal of Terramechanics*, Vol. 6, No. 1, 1969, pp. 23–35.

[2] M. Vukobratovic, A. A. Frank and D. Juricic, "On the Stability of Biped Locomotion," *IEEE Transactions on Biomedical Engineering*, Vol. BME-17, No. 1, January 1970, pp. 25–36.

[3] F. Gubina, H. Hemami and R. B. McGhee, "On the Dynamic Stability of Biped Locomotion," *IEEE Transactions on Biomedical Engineering*, Vol. BME-21, No. 2, March 1974, pp. 102–108.

[4] H. Hemami and B. F. Wyman, "Modeling and Control of Constrained Dynamic Systems with Application to Biped Locomotion in the Frontal Plane," *IEEE Transactions on Automatic Control*, Vol. AC-24, No. 4, August 1979, pp. 526–535.

[5] S-Y Oh and D. Orin, "Dynamic Computer Simulation of Multiple Closed-Chain Robotic Mechanisms," In *Proceedings 1986 IEEE International Conference on Robotics and Automation*, San Francisco, CA, April 7–10, 1986, pp. 15–20.

[6] D. E. Orin and S-Y Oh, "Control of Force Distribution in Robotic Mechanisms Containing Closed Kinematic Chains," *Transactions of ASME Journal of Dynamic Systems, Measurement, and Control*, Vol. 103, No. 2, June 1981, pp. 134–141.

[7] L. Shih, "Dynamic Modelling and Simulation of Mechanisms Consisting of Closed and Open Kinematic Chains with Compliance," PhD Thesis, University of Wisconsin-Madison, 1986.

[8] M. Kaneko, T. Tanie and M. N. M. Than, "A Control Algorithm for Hexapod Walking Machine Over Soft Ground," *IEEE Journal of Robotics and Automation*, Vol. 4, No. 3, June 1988, pp. 294–302.

[9] J. M. Hollerbach, "A Recursive Lagrangian Formulation of Manipulator Dynamics and a Comparative Study of Dynamics Formulation Complexity," *IEEE Transactions of Systems, Man, and Cybernetics*, Vol. SMC-10, No. 11, November 1980, pp. 730–736.

[10] J. Y. S. Luh, M. H. Walker and R. P. C. Paul, "On-Line Computational Scheme for Mechanical Manipulators," *ASME Journal of Dynamic Systems, Measurement and Control*, Vol. 102, No. 2, June 1980, pp. 69–76.

[11] T. R. Kane and D. A. Levinson, "The Use of Kane's Dynamical Equations in Robotics," *The International Journal of Robotics*, Vol. 2, No. 3, Fall 1983, pp. 3–21.

[12] H. Bremer, "On the Dynamics of Flexible Manipulators," In *Proceedings 1987 IEEE International Conference on Robotics and Automation*, pp. 1556–1560, Raleigh, NC, March 31–April 3, 1987.

[13] D. J. Manko and W. L. Whittaker, "Body Propulsion Model of an Orthogonal Legged Walker," In *Twenty First Annual Pittsburgh Conference on Modeling and Simulation*, Vol. 21, Part 5, pp. 2231–2237, Pittsburgh, PA, May 3–4, 1990.

[14] C. Lanczos, *The Variational Principles of Mechanics*, Univeristy of Toronto Press, Toronto, Canada, 2nd Edition, 1962.

[15] D. J. Manko and W. L. Whittaker, "Planar Abstraction of a Prototype Walking Machine," In *Twentieth Annual Pittsburgh Conference on Modeling and Simulation*, Vol. 20, Part 5, pp. 1817–1823, Pittsburgh, PA, May 4–5, 1989.

[16] H. Goldstein, *Classical Mechanics*, Addison-Wesley Publishing Company Inc., Reading, MA, 1950.

[17] J. W. Kamman and R. L. Huston, "Dynamics of Constrained Multibody Systems," *Transactions of ASME Journal of Applied Mechanics*, Vol. 51, December 1984, pp. 899–903.

[18] R. P. Singh and P. W. Likens, "Singular Value Decomposition for Constrained Dynamical Systems," *Transactions of ASME Journal of Applied Mechanics*, Vol. 52, December 1985, pp. 943–948.

[19] B. M. D. Wills, "The Load Sinkage Equation in Theory and Practice," In *Second International Conference ISTVS*, pp. 199–246, Quebec City, St. Jovite and Ontario, Canada, August 29–September 2, 1966.

[20] A. D. Sela and I. R. Ehrlich, "Load Support Capability of Flat Plates of Various Shapes in Soils," *Journal of Terramechanics*, Vol. 8, No. 3, 1972, pp. 39–69.

[21] R. Butterfield and M. Georgiadis, "Cyclic Plate Bearing Tests," *Journal of Terramechanics*, Vol. 17, No. 3, 1980, pp. 149–160.

[22] M. G. Bekker, *Theory of Land Locomotion*, University of Michigan Press, Ann Arbor, MI, 1956.

[23] M. G. Bekker, *Off-the-Road Locomotion*, University of Michigan Press, Ann Arbor, MI, 1960.

[24] M. G. Bekker, *Introduction to Terrain-Vehicle Systems*, University of Michigan Press, Ann Arbor, MI, 1969.

[25] J. Y. Wong, M. Garber, J. R. Radforth and J. T. Doxell, "Characterization of the Mechanical Properties of Muskeg with Special Reference to Vehicle Mobility," *Journal of Terramechanics*, Vol. 16, No. 4, 1979, pp. 163–180.

[26] E. J. Haug, S. C. Wu and S. M. Yang, "Dynamics of Mechanical Systems with Coulomb Friction, Stiction, Impact and Constraint-Addition-Deletion-I," *Mechanism and Machine Theory*, Vol. 21, No. 5, 1986, pp. 401–406.

[27] G. T. Rooney and P. Deravi, "Coulomb Friction in Mechanism Sliding Joints," *Mechanism and Machine Theory*, Vol. 17, No. 3, 1982, pp. 207–211.

[28] C. Canudas, K. J. Astrom and K. Braun, "Adaptive Friction Compensation in DC Motor Drives," In *Proceedings 1986 IEEE International Conference on Robotics and Automation*, Vol. 3, pp. 1556–1561, San Francisco, CA, April 7–10, 1986.

[29] A. Gogoussis and M. Donath, "Coulomb Friction Joint and Drive Effects in Robot Mechanisms," In *Proceedings 1987 IEEE International Conference on Robotics and Automation*, Vol. 2, pp. 828–836, Raleigh, NC, March 31–April 3, 1987.

[30] L. Petzold, "Differential Equations are not ODE's," *Siam Journal of Scientific and Statistical Computing*, Vol. 3, No. 3, 1982, pp. 367–384.

[31] R. P. Paul, *Robot Manipulators: Mathematics, Programming and Control*, MIT Press, Cambridge, MA, 1981.

[32] H. Asada and J.-J. E. Slotine, *Robot Analysis and Control*, John Wiley and Sons, New York, 1986.

[33] T. R. Kane and D. A. Levinson, *Dynamics: Theory and Applications*, McGraw-Hill Book Company, New York, 1985.

[34] D. J. Manko and W. L. Whittaker, "Inverse Dynamic Models of Closed-Chain Mechanisms with Contact Compliance," Accepted for Publication in the *ASME Journal of Mechanical Design*.

[35] F. Chorlton, *Textbook of Dynamics*, Halsted Press, New York, 1983.

[36] P. Lotstedt and L. Petzold, "Numerical Solution of Nonlinear Differential Equations with Algebraic Constraints I: Convergence Results for Backward Differentiation Formulas," *Mathematics of Computation*, Vol. 46, No. 174, 1986, pp. 491–516.

[37] L. R. Petzold, "Automatic Selection of Methods for Solving Stiff and Nonstiff Systems of Ordinary Differential Equations," Technical Report SAND80-8230, Sandia National Labratories Report, September 1980.

[38] A. C. Hindmarsh, "Numerical Solution of Ordinary Differential Equations: Lecture Notes," Technical Report UCID-16558, Lawrence Livermore Laboratory, June 1974.

[39] L. A. Hageman and D. M. Young, *Applied Iterative Methods*, Academic Press, New York, 1981.

[40] S. D. Chambers and V. M. Faires, *Analytic Mechanics*, The Macmillan Company, New York, 1943.

[41] R. A. Liston and E. Hegedus, "Dimensional Analysis of Load Sinkage Relationships in Soils and Snow," Technical Report 100, USATAC Land Locomotion Laboratory, Warren, MI, 1964.

[42] D. E. Freitag, "Soil Dynamics as Related to Traction and Transport Systems," In *Proceedings International Conference on Soil Dynamics*, Vol. 4, pp. 605–629, Auburn, AL, June 1985.

[43] C. W. Gear, *Numerical Initial Value Problems in Ordinary Differential Equations*, Prentice-Hall, Englewood Cliffs, NJ, 1971.

[44] W. H. Enright, *Numerical Methods for Systems of Initial Value Problems - The State of the Art*, Vol. F9 of *NATO ASI*, Springer-Verlag, Berlin, 1984. Computer Aided Analysis and Optimization of Mechanical System Dynamics, edited by E. J. Haug.

[45] O. Onafeko and A. R. Reece, "Soil Stresses and Deformations Beneath Rigid Wheels," *Journal of Terramechanics*, Vol. 4, No. 1, 1967, pp. 59–80.

[46] J. E. Bares and W. L. Whittaker, "Configuration of an Autonomous Robot for Mars Exploration," In *Proceedings 1989 World Conference on Robotics Research: The Next Five Years and Beyond*, Vol. 1, pp. 37–52, Gaithersburg, MD, May 7–11, 1989.

[47] J. E. Bares and W. L. Whittaker, "Configuration of a Circulating Gait Walking Robot," In *IEEE International Workshop on Intelligent Robots and Systems '90*, pp. 809–819, Tsuchiura, Japan, July 1990.

[48] S-M Song and K. J. Waldron, *Machines That Walk: The Adaptive Suspension Vehicle*, MIT Press, Cambridge, MA, 1989.

[49] S. Desa and B. Roth, *Mechanics: Kinematics and Dynamics*, John Wiley and Sons, New York, 1985. Recent Advances in Robotics, edited by G. Beni and S. Hackwood.

[50] W. H. Press, B. P. Flannery, S. A. Teukolsky and W. T. Vetterling, *Numerical Recipes: The Art of Scientific Computing*, Cambridge University Press, Cambridge, MA, 1986.

[51] R. R. Craig Jr., *Structural Dynamics*, John Wiley and Sons, New York, 1981.

[52] E. Krotkov, G. Roston and R. Simmons, "Integrated System for Single Leg Walking," Technical Report PRWP-89-3, Robotics Institute, Carnegie Mellon University, Pittsburgh, PA, 1989.

[53] H. J. Moore, R. E. Hutton, R. F. Scott, C. R. Spitzer and R. W. Shorthill, "Surface Materials of the Viking Landing Sites," *Journal of Geophysical Research*, Vol. 82, No. 28, 1977, pp. 4497–4523.

[54] R. Featherstone, *Robot Dynamic Algorithms*, Kluwer Academic Publishers, Boston MA, 1987.

[55] P. Nagy, "An Investigation of Walker/Terrain Interaction," PhD Thesis, Carnegie-Mellon University, 1991.

[56] P. Nagy, D. Manko, S. Desa and W. Whittaker, "Simulation of Postural Control for a Walking Robot," In *Proceedings 1991 IEEE International Conference on Systems Engineering*, pp 324–329, Dayton, OH, August 1–3, 1991.

Index

AMBLER
   circulating crawl gait 41
   configuration 39
   coordinate frames 42
   D-H parameters 43
   generalized forces
     body 44
     joint 45
   generalized coordinates 40
   global foot positions 43
   massless leg model 48, 55, 63, 68, 105
   model parameters 52
   physical parameters 45
   simulations
     body leveling 75
     body move 63
     flat settlement 54
     gait cycle 73
     leg recovery 68
     sloped settlement 59
   specialized equations 45

bilinear equation 11, 21, 97

closed-chains 5, 7–9, 17
constraint
   addition-deletion 14
   equations 6–8
   forces 6–9
   holonomic 9
   non-holonomic 8, 9
coordinate reduction
   Gaussian elimination 8, 9
   orthogonal transformation 10
   singular value decomposition 10
   zero-eigenvalues theorem 10
Coulomb's equation 13
   transition 13, 21, 23
coulomb friction 14, 27, 29

differential/algebraic equations 33, 37
dynamic formulation
   D'Alembert's Principle 8
   Kane's dynamics 8, 18
   Lagrangian 7, 18
   Newton-Euler 7, 18
   recursive Lagrangian 7, 18
   recursive Newton-Euler 6, 7

Euler methods
   explicit 35
   implicit 34, 37
   predictor-corrector 35

forward dynamic model 5, 6
foot-soil interactions 10, 20
   experiments 97
   lateral loading 21
   rotational loading 25
   sloped surfaces 25
   structural compliance 25
   vertical loading 21, 97

generalized coordinates 7, 17, 40

homogeneous transformations 7, 18

inverse dynamic model 6, 7
iterative methods
   chord 36, 70, 103
   convergence ratio 36, 103

Index

iteration error 36
Jacobian 36, 37, 103
Newton
  altered 36, 70, 103
  full 36, 70
  perturbation parameters 36, 37, 103
  residual vector 37

Jacobian 36, 37, 103
joint backdriving 14, 29, 31

Lagrange multipliers 7–9, 14
lateral soil loading 13
  brittle soils 13
  Coulomb's equation 13
  force-deflection equation 13
legged locomotion models
  biped 5
  compliant joints 6
  forward dynamic 5, 6
  inverse dynamic 6
  natural terrain
    boundary conditions 33, 37
    computational performance 70
    dynamic formulation 18
    equations of motion 19
    generalized coordinates 17
    generalized forces 19, 44, 45
    joint backdriving 29, 31
    joint damping 26, 29
    lateral foot displacements 22
    massless leg model 48, 55, 63, 68, 105
    solution procedures 33
    verification 51
  non-compliant 5
  static 6
  vertical foot-soil 6
local error 35

Overlapping Walker
  configuration 84
  coordinate frames 85
  D-H parameters 85
  physical parameters 84

simulations 86
Single Leg Testbed 26, 97
specialized equations 84

parallel processing 20
perturbation parameters 36, 37, 103

stepsize 35, 38
stiction 14, 27, 29
stiff
  differential equations 34, 63
  stability 34
structural compliance 6, 25

vertical soil loading 10
  analytical method 11
  dimensional analysis 11
  force-deflection equations
    asymptotic 10
    bilinear 10
    exponential 10
    modified power 10
    power 10
  non-homogeneous soils 11
  repetitive loading 11
viscous damping 14

UNIVERSITY OF STRATHCLYDE

30125 00491341 3

Books are to be returned on or before the last date below.

11 JAN 1996

- 6 MAR 2001

04 APR 1996

- 5 JAN 1999

29 JAN 1999